U0175261

经典面食文化

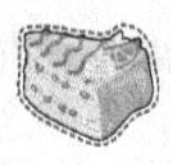

曹竑　马忠仁　主编

科学出版社

北　京

内 容 简 介

本书精选了朝鲜冷面、包子、饺子、金华酥饼、萨其马、馓子、山西刀削面、山西花馍、凉皮、兰州牛肉面、新疆烤馕、印度飞饼、俄罗斯黑面包、皮塔饼、匈牙利姜饼、派等经典面制品，比较系统地阐述了各种面制品的起源及发展、逸闻趣事、原辅料介绍、制作工艺流程、操作要点及风味特点等。

全书内容丰富，图片精美，融文化性、实用性于一体，文字精练，以英文为导引，以美食为桥梁，促进各国文化交流，雅俗共赏。可供面制品生产从业人员、大专院校学生使用，也可供广大读者参阅。

图书在版编目（CIP）数据

经典面食文化 / 曹竑，马忠仁主编．—北京：科学出版社，2020.6
ISBN 978-7-03-065370-3

Ⅰ．①经…　Ⅱ．①曹…②马…　Ⅲ．①面食—文化　Ⅳ．①TS972.132

中国版本图书馆CIP数据核字（2020）第093474号

责任编辑：席　慧　林梦阳 / 责任校对：严　娜
责任印制：徐晓晨 / 封面设计：蓝正设计

科学出版社 出版
北京东黄城根北街16号
邮政编码：100717
http://www.sciencep.com
北京建宏印刷有限公司 印刷
科学出版社发行　各地新华书店经销
*
2020年6月第　一　版　开本：787×1092　1/16
2024年1月第三次印刷　印张：9
字数：186 000
定价：69.80元
（如有印装质量问题，我社负责调换）

《经典面食文化》编写委员会

主　　编　曹　竑　马忠仁

副 主 编　丁功涛　田晓静

编写人员　（按姓氏拼音排序）

曹　竑　丁功涛　高丹丹　马忠仁

田晓静　魏　嘉　张福梅　周雪雁

项 目 资 助

1. 科技部科技伙伴计划，中国 - 马来西亚清真食品国家联合实验室建设项目（KY201501005）

2. 西北民族大学生物工程“双一流”和特色发展引导专项经费（No.1001070204）

前言

PREFACE

民以食为天，食以安为先。本书针对“经典面制品”的文化背景、历史故事、现实演变、烹饪方法等进行系统介绍，旨在增进人们对饮食文化的了解及对面食文化的认知，以美食为契机促进民族团结和对外友好交流。

面制品历史悠久，深受世界各国人民的喜爱。为了挖掘、整理和发扬各式面制品的制作技术，总结、推广和交流各式面制品的制作经验，研究、探讨和提高产品的技术性、艺术性和科学性，使其不断创新、不断发展，我们尝试编写了这本《经典面食文化》。

本书共介绍了16种经典面制品，包括朝鲜冷面、包子、饺子、金华酥饼、萨其马、馓子、山西刀削面、山西花馍、凉皮、兰州牛肉面、新疆烤馕、印度飞饼、俄罗斯黑面包、皮塔饼、匈牙利姜饼、派等。在编写内容和形式上，每种面制品按起源及发展、逸闻趣事、原辅料介绍、制作工艺流程、操作要点及风味特点的顺序做了详细介绍，以便查阅和参考。

参加本书编写的有：曹竑（兰州牛肉面、新疆烤馕），丁功涛、马忠仁（匈牙利姜饼、印度飞饼），田晓静（皮塔饼、凉皮、萨其马），高丹丹（饺子、馓子），魏嘉（金华酥饼、俄罗斯黑面包、派），张福梅（包子、朝鲜冷面、山西花馍），周雪雁（山西刀削面）；马忠仁、丁功涛、马莹萍负责英文翻译；曹竑、田晓静、丁功涛负责确定全书编写计划、约稿和审稿。

本书内容丰富，重知识性和趣味性，文字精练，语言朴实生动，融文化性、实用性于一体，全彩印刷，版式设计精美。书中全方位介绍了经典面食的制作工艺、背后的故事和当地的饮食习惯和风俗，突出了民族食品文化主题。每种面食都配有精美图片，且有英文导引，旨在以美食为桥梁，促进各国文化交流，雅俗共赏。

本书承蒙科技部科技伙伴计划——中国 - 马来西亚清真食品国家联合实验室建设项目 (KY201501005) 及西北民族大学生物工程“双一流”和特色发展引导专项经费（No.1001070204) 资助，并得到甘肃省食品工业协会成富山会长的热情指导，凝聚了全体编审人员的辛勤汗水。刘元林、龙鸣、马石霞、徐珊、张娅莉、李荣等同学在资料收集、图表整理过程中给予很大帮助，部分照片由火爱萍、李临菊提供，同时书籍编写过程中还参考了大量资料，向这些资料的作者致以诚挚的感谢。

本书内容涉及面广，由于作者水平所限，不足之处在所难免，恳请广大读者批评指正。

编者

2020年5月于西北民族大学

目录

CONTENTS

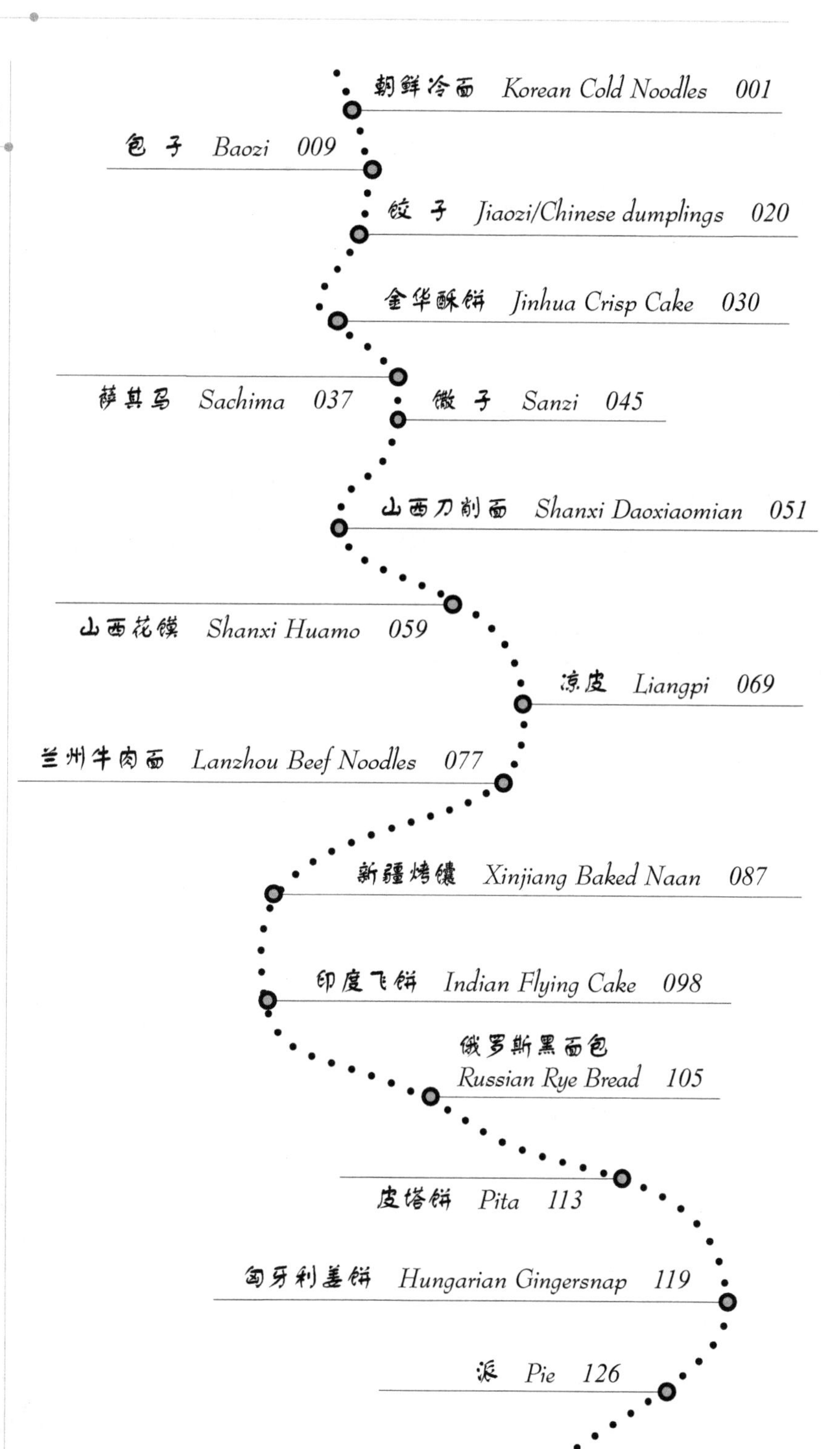

朝鲜冷面

Korean Cold Noodles

Korean cold noodles, also named as "Goryeo Noodles", are made from high-quality buckwheat flour and potato starch. They are known as the "cold noodles" for that they are not eaten until they cool down after cooked on fire. As one of the three specialties of the Korean ethnic group, Korean cold noodles are well-known both at home and abroad for their fine craftsmanship and unique dietary flavors.

Originated from the Korean Goryeo Dynasty, the cold noodles have a history of more than 200 years. Korean cold noodles, which dates back to around 150 years ago, were initially introduced into China from the end of the 19th century with the migration of the Korean people. Being the most representative and popular traditional Korean food, it is well-known throughout the world.

Korean cold noodles have a long history and unique recipe. They have been invented and developed by the Korean people during their long-term production and life experience with distinct national and regional characteristics. They not only have the features of Korean culture, but also have the value for historical, folk culture and academic researches.

The main ingredients are buckwheat flour, wheat flour, sorghum rice flour and corn flour, and also starch. Generally the Korean cold noodles are made of buckwheat flour with beef soup or chicken soup, added with spicy cabbage, sliced meat, eggs, shredded cucumber and pear, etc. First, put a small amount of cold soup and some noodles in the bowl, then add the seasoning, and finally pour the soup again.

The noodles are thin and a bit chewy, and the soup tastes cool, sour and spicy. No matter in hot summer or cold winter, the Korean cold noodles would be the Koreans' first choice when they want noodles.

The cold noodles on the menu currently have been greatly improved and adjusted on the basis of the original. Beef soup, onion, garlic, and other vegetables and fruit juices(such as pineapple juice)are included to add the flavor of the soup. The flavor of the soup determines the overall taste of the cold noodle and it can indicate the cooking capacity of a chef.

Korean cold noodles are a highlight within the food industry. Once they appear on the menu, they boast spreading popularity and have certain influences in the world. Based on the advantages of fast food delivery, small investment, labor saving and low cost of ingredients, the cold noodles have a broad market prospect.

Since ancient times, the Korean people have had the tradition of eating cold noodles at noon of the fourth day of the first lunar month or on their birthday. On this day, eating thin and long cold noodles represents one will live a long life. Therefore, cold noodles are also called "longevity noodles". There are many touching legends about Korean cold noodles. These legends add an air of romance to Korean cold noodles. The Korean people has the custom of eating cold noodle on the fourth day of the first lunar month for legend said that eating cold noodles on that day can enjoy longevity. Therefore, eating the "longevity noodles" on the fourth day of the first lunar month has become the traditional custom of the Korean nationality.

Korean cold noodles are very popular at home and abroad due to the unique flavor. They have three characteristics: first, the features of Korean eating customs; second, the fine making process; finally, the unique flavor. The Korean cold noodles are generally served with bowls and also plates. The cold noodles are famous for the sour, spicy, and sweet flavor. Yanji cold noodles include chili sauce and less soup; Pyongyang cold noodles have more soup and light flavor.

The cold noodles made of wheat flour are pale yellow or yellow; of buckwheat flour, brownish yellow, brown, or tan; of corn flour and other miscellaneous grains, yellow, golden yellow, or in other miscellaneous grain colors. Korean cold noodles have the taste and smell inherent in wheat flour(buckwheat flour, cornmeal, or other miscellaneous grains) without unpleasant odor. They have uniform thickness, smooth surface, and transparency without white knots or bubbles. There is no muddy soup after they are reconstituted. They are not sticky but soft and chewy. They have such palatability as being chewy, tender, refreshing, and smooth and emit the noodle aroma(scent of buckwheat noodles, corn or other miscellaneous grains).

朝鲜冷面又叫“高丽面”，是选用优质荞麦面和土豆淀粉为原料压成的面条。这种风味小吃热制冷吃，故称“冷面”。朝鲜冷面被誉为朝鲜族三大特色食品之一，以其精细的制作工艺和独特的饮食风味而享誉国内外。

一 朝鲜冷面的起源及发展

冷面起源于高丽朝时期。朝鲜冷面，是从19世纪末随朝鲜人迁入而传入中国，距今约有150年。其历史悠久，工艺独特，是朝鲜族人民在长期的生产生活中创造并不断丰富和发展起来的，具有鲜明的民族特色和地区特点。

传统朝鲜冷面是以小麦粉、淀粉、荞麦粉为主要原料，添加或不添加辅助配料（食用碱、食用盐和面条品质改良剂），采用挤压工艺加工生产的具有一定熟化度和特殊口味的冷面条。现今市售的冷面已经在原先的基础上做出了相当多的改进和改良，调味中有牛肉原汁，又有洋葱、蒜头之类的香料，加入了其他口味的各式蔬菜汁及水果汁（如菠萝汁）以提升汤汁的口感。在荞麦冷面深受人们青睐的同时，又相继而生了一系列以玉米面、高粱面、黑米粉、红豆粉、绿豆粉等杂粮为原料的新式冷面，冷面家族系列产品不断扩大，逐渐流行全国，食用人群和方法远远超出了传统意义上的冷面领域。朝鲜冷面凭借出餐快、投资小、节约人工、材料成本低几大优势，具有广阔的市场前景。

二 朝鲜冷面的逸闻趣事

（一）赞美诗

黑龙江诗人谢晓光写七绝《冷面辣菜》赞美朝鲜冷面。

白山冷面滋人美，黑水调羹辣菜香。
东北游侠轻富贵，延边乡味忆悠长。

（二）冷面的传说

关于朝鲜族冷面还有很多美丽的传说，这为朝鲜族冷面添加上一层传奇色彩。朝鲜族有在农历正月初四中午吃冷面的习俗，说是如果这一天吃了冷面，就会像长长的面条那样长寿。因此，冷面又被称作“长寿面”。每年农历正月初四吃“长寿面”就成了朝鲜族的传统习俗。

三 朝鲜冷面的经典制作工艺

传统朝鲜冷面一般是即食冷面，由面条、冷面汤、冷面帽、点缀、冷面酱五个部

分组成。冷面的面条由荞麦面加一定比例的淀粉或面粉混合和面，压成圆面条，直接煮熟后浸以冷水冷却而制成；冷面汤由牛肉或牛骨熬制成汤后调制而得；冷面帽和点缀是放在面条上的辅料，起到增进营养、口味、感官等效果，主要有牛肉片、熟鸡蛋、蔬菜丝、泡菜类、水果片等；冷面酱由辣椒粉、姜汁、蒜汁、食盐、熟白芝麻等调制而成。

（一）原辅料介绍

朝鲜冷面的原料主要有荞麦粉、淀粉、高粱米粉和玉米粉四种，其中尤以荞麦粉冷面著称，还可用榆树皮面和淀粉制作。一般用牛肉汤或鸡汤，佐以辣白菜、肉片、鸡蛋、黄瓜丝、梨条等。食用时，先在碗内放少量凉汤与适量面条，再放入佐料，最后再次浇汤。其面条细质韧，汤汁凉爽，酸辣适口。从制作方式上看，冷面更像是辣白菜的衍生产品，因为更原始的冷面汤就是家庭腌白菜的菜汤，顶多放进些炖牛肉时剩下的汤汁，其特有的酸、辣、鲜、香是天然发酵而成的，并少有人工的调味。

朝鲜冷面之主料——荞麦粉

荞麦因其含有丰富的营养和特殊的健康成分而颇受推崇，被誉为健康主食品。荞麦粉是制作朝鲜冷面的主要原料，可赋予朝鲜冷面独特的风味和颜色。

朝鲜冷面之主料——淀粉

淀粉也是用来制作朝鲜冷面的主要原料，主要有绿豆淀粉、木薯淀粉、甘薯淀粉、红薯淀粉、土豆淀粉、麦类淀粉、菱角淀粉、藕淀粉、玉米淀粉等。淀粉不溶于水，在和水加热至60℃左右时，则糊化成胶体溶液。朝鲜冷面制作时淀粉要选用色泽洁白、糊化温度低（59～69℃即可糊化）、糊化速度快、黏性大、透明度高的土豆淀粉。淀粉可赋予朝鲜冷面良好的质地和口感。

朝鲜冷面之辅料——玉米粉、高粱粉

玉米粉维生素含量非常高，钙含量接近乳制品，含有微量元素硒和镁；高粱粉蛋白质中赖氨酸含量较低，属于半完全蛋白质。玉米粉、高粱粉可赋予朝鲜冷面良好的风味，且其本身的营养成分还可增加朝鲜冷面的营养价值。

朝鲜冷面之佐料

辣白菜、肉片、鸡蛋、黄瓜丝、梨条、葱丝、辣椒、芝麻、蒜泥、味精、盐、醋、酱油等。

朝鲜冷面风味的灵魂——冷面汤料

朝鲜冷面的汤可分为荤汤和清汤。荤汤一般用牛骨头、牛肉的清汤或鸡汤；清汤一般用米白醋、酱油、盐、糖、味精、少许姜末、蒜末、香菜、黄瓜丝调制而成，还可加入梨片、苹果丝、菠萝、胡萝卜等，不仅有酸、甜、咸、辣、鲜味及营养搭配，还无牛肉膻味，不油腻，清爽可口。配制的冷面汤一般冷却成－4～0℃，泡入面条，柔韧耐嚼，酸酸甜甜，清凉爽口，滑顺润喉，令人食欲大增。

目前市场上流通的冷面汤料有两种，一种是粉状，是由辣椒粉、食盐、白糖、

酸味剂等简单地混合而制成，营养不够丰富；另一种是液体状，是浓缩的牛肉汤和食盐、白糖、醋等混合加工而成，味道鲜美可口，但是不适合用于方便冷面的生产和流通。

朝鲜族冷面的汤料不仅要做到酸、甜、咸、辣、鲜五味俱全，而且还要无膻味、不油腻、清爽可口，同时还富含营养价值。

（二）朝鲜冷面的制作

朝鲜冷面制作工艺流程

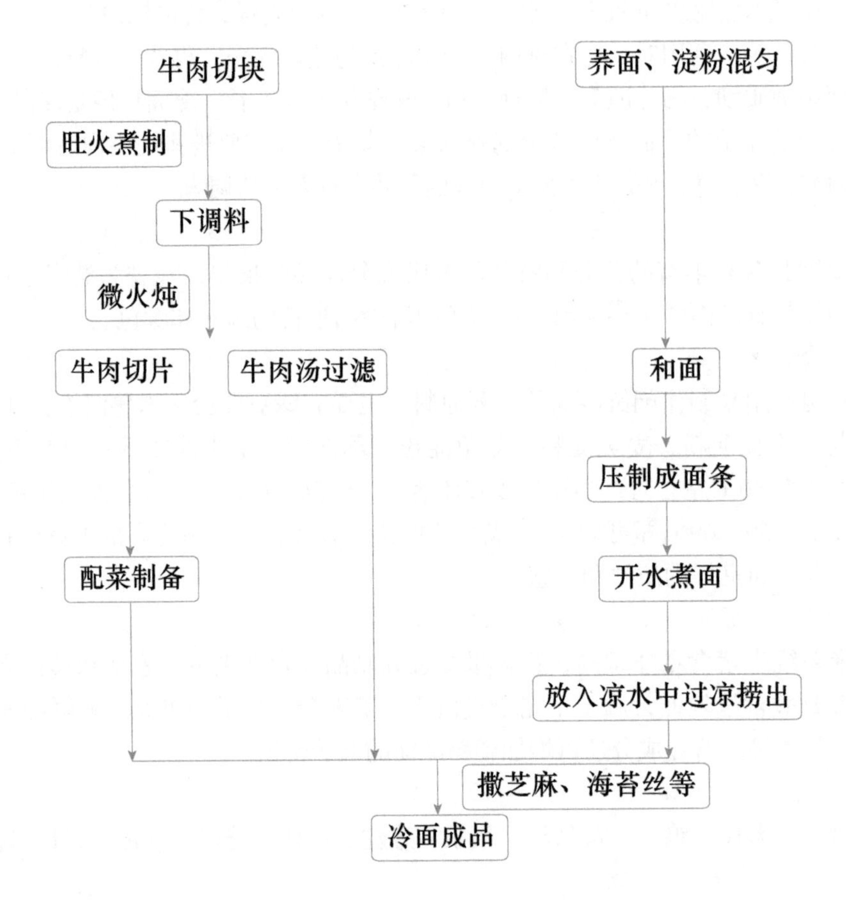

参考配方

荞麦面粉800g，土豆淀粉200g，精牛肉1000g，白菜2000g，胡萝卜200g、黄瓜200g、苹果、梨各300g，鸡蛋丝100g，熟芝麻20g，辣椒面、蒜泥、姜末、酱油、醋、香油、海苔丝、松仁、食碱、精盐、味精各少许。

操作要点

1．汤的制作：选择新鲜的牛肉，称取1000g切成适宜大小（200g），用清水浸泡30min去血水；取5kg冷水，加入牛肉用旺火边煮边撇出血沫，熬煮40min，捞

出牛肉，立即冷却汤，并撇去浮层牛脂，用四层纱布过滤得牛肉清汤；葱、姜、蒜洗净去除不可食用部分；胡萝卜、黄瓜要新鲜无糠心，洗净去皮，切丝；将苹果和梨去把、削皮、去核、切成薄片；红枣清洗干净。将这些辅料称量好后用纱布包好，与牛肉、牛肉清汤再次煮沸 20min，去除牛肉汤的膻味和不良气味，增强滋味和香气，得到冷面汤基料；待牛肉完全炖熟时，捞出放置案板上，等其晾凉后切成薄片（4.5cm×3cm）待用，将牛肉汤稍过滤后冷却；称好食盐、白糖等辅料添加到冷面汤基料中调味。完全溶解混合均匀，得到最终产品，放入容器内待用。

2．辣白菜的制作：将白菜择去青叶洗干净，沥净水分，切成 3cm 长的白菜丝，加入精盐 50g、辣椒面 30g、蒜泥 50g、姜末 6g，搅拌均匀，装入瓷缸内自然发酵（保持温度在 15℃左右）2 天，即成辣白菜。

3．调味酱的调制：辣椒面 50g、蒜泥 20g，用 100g 冷水调匀，捶成粥状，即成调味酱。

4．面团的制作：用 50g 温水溶化食碱，把荞麦面和土豆淀粉搅和均匀，加入 1000g 开水和成烫面，宜硬不宜软，防止粘成团。

5．冷面的制作：将面团揉好，揉成圆条，放入特制的挤筒内，开动冷面机，压制成条；煮面，煮至面条表面呈现出光泽时即熟（40～60s），用大漏勺捞出面条，放入箩筐内，用冷水投洗数遍，沥干水分至无黏液为止；将面条放在案板上拉断成约为 45cm 长的条，捋成团状放入碗中央；按顺序放胡萝卜丝、黄瓜丝、苹果片、梨片，淋上韩式辣酱，放入辣白菜，依次堆成塔状；五片熟牛肉、半个水煮蛋、少许蛋皮丝放在侧边排开；撒芝麻、海苔丝、白糖少许、香油、松仁，最后浇上牛肉汤，滴上香油即成。

四　朝鲜冷面的风味特色

朝鲜冷面以其独特风味享誉国内外，它具有朝鲜族饮食习俗特点，制作工艺精细，风味独特。朝鲜冷面口感清凉爽滑、筋道，甜、酸、辛、辣、香五味俱全，尤以酸辣、酸甜风味著称。其中，延吉冷面加辣酱，汤水少；平壤冷面汤水多，口味清淡。

朝鲜冷面中小麦粉冷面呈浅黄色或黄色等，荞麦粉冷面呈棕黄色、棕色或棕褐色等，玉米面及其他杂粮冷面呈黄色、金黄色或呈其他杂粮颜色。朝鲜冷面具有小麦粉（荞麦粉、玉米面或其他杂粮）固有的滋味和气味，无异味；粗细均匀，表面光洁，无白结、气泡，有透明感；复水后不浑汤，口感不粘牙，柔软有咬劲，具有筋、软、爽、滑适口性及面香味（荞面香、玉米或其他杂粮香味）。

经典朝鲜冷面以荞面为主要原料制作而成。荞麦粉营养丰富，含有丰富的赖氨酸，铁、锰、锌等微量元素比一般谷物丰富，而且含有丰富的膳食纤维、烟酸和芦丁，对降低血脂和胆固醇、软化血管、保护视力和预防脑血管出血有帮助。

参考文献

范洙，李东玲，赵金伟，等. 2009. 朝鲜族传统冷面汤配方的优化研究 [J]. 中国调味品，34（09）：72-76，79.

李京一. 2009. 从延吉冷面到韩国烧烤 [J]. 北京观察，（09）：17-18.

卢敏，殷涌光. 2005. 荞麦粉的添加率对朝鲜族冷面品质的影响 [J]. 河南工业大学学报（自然科学版），（03）：34-36.

卢敏，殷涌光. 2005. 影响朝鲜族冷面品质的原料因素研究 [J]. 粮食与饲料工业，（09）：29-30.

王成军，李勇. 2005. 方便朝鲜冷面加工技术 [J]. 食品工业，（05）：16-17.

辛若竹，郑贤光，吴鸣，等. 2013. 吉林省冷面产品质量技术规范的研究 [J]. 中国卫生检验杂志，23（04）：979-983.

徐祥鹏. 2002. 朝鲜族泡菜 · 冷面 · 打糕 [J]. 四川烹饪高等专科学校学报，（04）：35.

余平，冷进松. 2010. 荞麦粉膨化工艺参数的优化及其在朝鲜族冷面加工中的应用 [J]. 粮食加工，35（01）：79-82.

张美莉. 2009. 怎样做朝鲜冷面 [J]. 农家顾问，（11）：54.

张美莉. 2010. 朝鲜冷面与苦荞速食面的制作 [J]. 农村新技术，（02）：51-52.

包 子

Baozi

Baozi is a traditional Chinese delicacy. Its wrapper is made from flour, water and sugar and its stuffing contains meat, vegetable, fine bean mash, etc. It should be made with the process of fermentation and steaming.

According to the legend, Baozi was invented by Zhuge Liang in the Three Kingdoms period, and he was regarded as the founder of dough modelling. In the northern China, steamed food without stuffing is generally called Mantou, and those with stuffing are called Baozi. Instead, the southerners would like to call the steamed food with stuffing as Mantou, and that without stuffing as Big Baozi.

During the Tang and Song Dynasty, Mantou gradually became the main course among the rich families. Besides Han people in the central China, the noble Khitay tribe of Liao Kingdom enjoyed eating Baozi. The use of the name "baozi" began in the Song Dynasty.

Mantou and Baozi had a clear distinction in Qing Dynasty. The book *Qing Bai Lei Chao* (a compilation of various anecdotes in Qing Dynasty) documented that Mantou, once called Manshou, was made by fermented flour and steamed into the round shape.

Nowadays, the northerners and the southerners still keep different name for Baozi and have various fillings and tastes of Baozi. Northerners have typical character of frankness and they have wide choices about the filling of Baozi, such as carrot, seaweed, vermicelli, egg, eggplant, dried tofu and pickle. Baozi with such fillings tastes crisp or soft. While southerners have sensitive and thoughtful character and they prefer light and mild flavor. Therefore, they seek for the exquisite making of Baozi. The flour wrapper should be thin but refined enough to hold the fillings, which can show the excellence of skills for making Baozi.

There are also some anecdotes about Baozi. For example, "Goubuli" is originated from the year 1858. During the reign of the Emperor Xianfeng in Qing Dynasty, there was a young man named Gao Guiyou in Wuqing county (now Wuqing district, Tianjin city). As his father had this son at the age of 40, the boy was named as "Gou Zi" (nickname), meaning hopes for his happiness and safety. As Gao Guiyou was highly-skilled in making Baozi, the Baozi he made tasted soft, fresh and not greasy. With its color, fragrance and shape being unique, the Baozi attracted lots of people coming from afar to eat. His business became very prosperous and his Baozi shop enjoyed good reputation. As more and more customers came to eat, Gao Guiyou were too busy to talk to them. Thus, the customers jokingly said "when Gou Zi sold the Baozi, he totally ignored others". As the time went by, people called him as "Goubuli" and called the Baozi as "Goubuli Baozi" while the original name of the shop was gradually forgotten. It was said that when Yuan Shikai was appointed as the governor to train the new army in Tianjin city, he once presented "Goubuli Baozi" as a tribute to Empress Dowager Cixi in Beijing. Empress Dowager Cixi said with joyfulness after taste, "any animal meat or seafood are not as delicious as the Goubuli Baozi, which can make people have a long live." Since then, Goubuli Baozi has gained great popularity and chain stores were opened in many places.

In addition, Qingfeng Baozi shop was initially built only as a small ordinary restaurant in 1948. Due to the delicious flavor of its Baozi, it began to sell Baozi only since 1956 and officially set the brand "Qingfeng Baozi shop". Currently, the Qingfeng Baozi has become a special snack in Beijing, and the old brand Qingfeng Baozi shop has been gradually introduced into international fast food market. After President Xi's visit, Qingfeng Baozi shop nowadays has become a new tourist destination. The Qingfeng Baozi shop in Yuetan North Street is always full of customers waiting for taking pictures.

Baozi tastes delicious and fresh as it is made from quality raw materials. The surface of Baozi looks half-transparent and glossy with organized figures on it. Baozi can be regarded as both staple food and non-staple food. It includes both meat and vegetables, and contains a variety of nutrients, in line with the requirements of a balanced diet.

包子是一种将面粉加水、糖等调匀并发酵，制作时加入肉、菜、豆蓉等馅料，通过蒸笼蒸制或烤制而成的食品，是中华民族的传统美食。

一　包子的起源及发展

包子是中国汉族传统食品之一，相传由三国时期诸葛亮发明，诸葛亮也被尊奉为面塑行的祖师爷。“包子”这个名称的使用始于宋代，《燕翼诒谋录》：“宫中出包子，以赐臣下”。馒头之有馅者，北方人谓之包子。北方一般将无馅的蒸食称为馒头，有馅者称为包子，而南方人则一般称有馅者为馒头，无馅者为“大包子”。

唐、宋年间馒头已作为美馔，是殷富人家的主食。中原汉族人家喜欢食用，而且它还是辽国契丹贵族的食品。从文物考古中发现的一座辽墓壁画上，有侍女端着一盘馒头送给主人进餐，画中反映出馒头已成为契丹族家庭的膳食珍品。

宋代著名的大诗人陆游《蔬园杂咏・巢》：“昏昏雾雨暗衡茅，儿女随宜治酒殽，便觉此身如在蜀，一盘笼饼是豌巢。”自注：“蜀中杂彘（猪）肉作巢馒头，佳甚。”

到了清代，馒头和包子才有明确的区分。清末民初徐珂《清稗类钞》辨馒头：“馒头，一曰馒首，屑面发酵，蒸熟隆起成圆形者。无馅，食时必以肴佐之。”“南方之所谓馒头者，亦屑面发酵蒸熟，隆起成圆形，然实为包子。”

如今，北方和南方仍保留着对包子的不同叫法，而且馅料与口味也有差别。北方人性格豪爽，在包子馅料的选择上更为粗犷，胡萝卜、海带、粉丝、鸡蛋、茄子、豆干、酸菜等均可入馅，口感或酥烂，或生脆。南方人细腻温和，崇尚清淡、量小的饮食习惯，似乎不太重视饱腹感，而更追求精细的制作水准，要求皮薄却不漏汁，显示制作技艺之高超。

包子行业历经千余年历史经久不衰，凭借其风味独特、营养均衡、服务快捷，有着广泛的群众基础和巨大的市场空间。

二　包子的逸闻趣事

（一）“一条龙”包子

南北朝时期（420 年～589 年）的陈国建都在建康（今江苏南京），末代皇帝陈叔宝（553 年～604 年）人称“陈后主”。陈叔宝生长于深宫之中，不知稼穑之艰难。有一天他待在宫中闷得慌，就穿了一身便服，悄悄地溜出了宫门，跑到秦淮河边转悠。那里店铺、茶楼、布庄、米号等门面挨门面，多得数不清。金粉楼台，鳞次栉比；画舫凌波，桨声灯影，构成一幅如梦如幻的美景奇观，为文人墨客聚会的胜地。他挤在人群中，左瞧瞧右看看，看得眼花缭乱，觉得一切都很新奇好玩。走着玩着，

不知不觉来到一家包子铺门前，才出笼的包子，香气四溢，馋得他口水直流。他想吃，可又不懂花钱买的规矩，便伸手拿起一只刚出笼的包子，张嘴就咬，越吃越觉得包子味道好。吃了一个又一个，撑得肚皮滚圆，一抹嘴就走了。店家看他穿着绸缎，不像寻常人家子弟，也没敢吱声。过了几天，陈后主又溜出宫来拿包子吃。店主忍不住了，说："小主顾，这铺面本小利微，入不敷出，您来吃包子，一次两次算小的请客，天天来吃，实在消受不起。瞧您穿戴是大户人家，不在乎几个小钱，还是请留个账头，日后也好侍候。"陈后主听到此话觉得奇怪，他可是从来没有听说吃东西还要给钱，只好瞪着眼，不知如何是好。不过店家要他留名，他倒听懂了，心想：我的名儿谁敢叫！于是随口就说："朕是一条龙。"这倒让店家听不明白，只好递过笔，让他留下大名。陈后主抓起笔，歪歪扭扭地写了"一条龙"三个字。后来人们知道那叫"一条龙"的小孩，就是小皇帝陈后主。达官贵人蜂拥而至来吃包子，包子铺生意也更加兴旺。包子铺门口曾是皇帝站过的，人们就叫它"龙门"，那条街被称作"龙门街"，"一条龙"三个字也被装裱上了中堂。"一条龙"的包子名声远扬，一直流传至今。

（二）"狗不理包子"

"狗不理"包子的始祖高贵友的父母是贫苦农民，清朝道光年间全家逃荒到直隶武清县下朱庄（现天津市武清区）居住。1831年（清道光十一年），其父四十岁时得子，大名高贵友，为求平安，取其乳名"狗子"，期望他能像小狗一样好养活。高贵友从小性格倔强，他的父亲害怕他在村子里惹是生非，15岁就托人把他带出去学手艺，找事做。恰好坐落于天津侯家后街的刘家蒸吃铺需要伙计，高贵友就被介绍了进去。刘家蒸吃铺坐落于天津南运河边上，主要经营蒸食和肉包，高贵友名为学徒，实际是干杂活。因能吃苦，干活勤快，店里的师傅们都很喜欢他，再加上人聪明伶俐，脑瓜子灵活，学东西很快，不久就学会做包子。

后来，刘记停业了，高贵友就在侯家后街搭起一个小棚，自己做包子卖。他为人忠厚老实，手艺好，做事又十分认真，从不掺假，做的包子真材实料，个大味鲜，深得百姓青睐。但小本生意，他雇不起伙计，只能一个人连做带卖，还要吆喝应声，忙得喘不过气来，就想了个法子，在摊头摆个碗儿，顾客把铜钱放在碗里递过来，他就按钱给包子，一言不发。于是"狗子卖包子，一概不理"这话一传十十传百，包子便被叫作"狗不理包子"。等到日后，高贵友在棚子附近买下了一间小门面，取名"德聚号包子铺"。因他做出来的包子特别好吃，名声很快就响了起来，久而久之，人们习惯称它"狗不理包子铺"，原店铺字号却渐渐被淡忘。

高贵友的包子不同凡响，包子褶花匀称，每个包子都是18个褶。口感柔软，鲜香不腻，清香适口，形似菊花，爽眼舒心，色香味形都独具特色。还有人作《西江月》赞美"狗不理包子"："一体雪白肌肤，小巧玲珑身材，更喜香腮绽笑靥，凝睇越看越爱。一朝取之入口，满嘴清香爽快。蟠桃宴上一美味，如今降落尘埃。"据说，慈禧太后品尝“狗不理包子”后赞道：“山中走兽云中雁，陆地牛羊海底鲜，不及狗不理香矣，食之长寿也。”从此，狗不理包子越传越有名，由小摊档变成大摊档，由大摊档变成包子铺，变成饭馆，又扩大成三联号。不仅在国内一些城市建了分店，还出口到了日本，走向世界，进入许多国家市场，倍受宾客欢迎。

（三）庆丰包子

“庆丰包子铺”创建于1948年，原地址位于北京市西单东南角。起初是一家普通的小饭馆，只因所营包子口味地道，1956年公私合营后，专一经营包子。1976年正式更名为“庆丰包子铺”，以经营包子、炒肝为主，成为北京市的特味小吃，是北京市百姓认可的著名快餐品牌。2013年12月，习近平主席光顾北京老字号庆丰包子铺，自己排队买单、自己端盘找座，二两猪肉大葱包子、一份拌芥菜和一份炒肝的21元“主席套餐”从此就卖火了，庆丰包子铺也成了人们争相观赏拍照留念的一个新景点。

三 包子分类

包子按照馅料的种类分为肉包子和素包子；按制作方法分为蒸包子、烤（煎）包子、灌汤包子；按照地域分为天津狗不理包子、上海小笼包、北京庆丰包子、武汉鱼香包、成都韩包子、广东叉烧包、新疆烤包子、巴塘团结包子；按包子在造型上的变化又可分为光头包（奶黄包、蛋黄莲茸包）、提褶包（小笼包、鲜肉大包）、花式包（秋叶包、佛手包、葫芦包）；按馅口味分甜馅包子（如豆包、果馅包）、咸馅包子（肉馅包子及素馅包子）等。此外，国外比较著名的包子有俄罗斯油煎包子及韩国哈巴狗包子等。

四 包子的制作工艺

（一）原辅料介绍

包子的常用原辅料为面粉、猪五花肉、肉皮冻、酱油、油、羊肉、牛肉、粉条、香菇、豆沙、芹菜、包菜、韭菜、豆腐、木耳、梅干菜、蛋黄、芝麻、料酒、香油、白糖、葱花、姜末、精盐、胡椒粉、味精等。

包子皮制作原料——面粉

面粉是小麦经过清理除杂、润麦、研磨、筛粉等工艺制得的，按照其蛋白质的含量分为高筋粉（≥11.5%）、中筋粉（8.5%～11.5%）和低筋粉（≤8.5%）。面粉含有蛋白质、碳水化合物、灰分、酶、水分、脂肪和维生素等，是制作包子皮的主体材料，是形成面团的组织结构，提供酵母发酵所需的能量。

包子皮制作一般选用中筋粉，有时也选用高筋粉，要避免使用低筋面粉如蛋糕粉和避免使用玉米粉。

包子皮制作辅料——酵母

酵母是制作发面包子皮必不可少的一种重要生物膨松剂，常见的酵母有新鲜酵母、干酵母及快速酵母等几种。在包子皮制作中的功能主要是生物膨松、面筋扩展、风味改善及增加营养价值等。

包子馅制作原料——肉类

肉品种非常多，包括猪肉、羊肉、牛肉、鸡肉等，鲜肉绞碎加入调味料和蔬菜可制成鲜肉馅；经过加工后的肉品可制成如叉烧馅、酱肉馅、火腿馅等肉馅。肉类可增加包子风味、改善营养。

包子馅制作原料——蔬菜类

常用的蔬菜有韭菜、芹菜、萝卜、白菜、茴香菜、豆角、萝卜缨等，梅干菜泡发后也是非常好的馅料。蔬菜制作包子馅，清素爽口，热量低，有一定的保健作用。

包子馅制作调味料——料酒、香油、白糖、葱花、姜末、精盐、胡椒粉、味精、食盐等。

酱油可使包子馅入味，增加色泽；料酒可使肉馅去除腥味。

糖是一种富有能量的甜味料，是酵母的主要能量来源，能改善面团的物理性质及面团内部的组织结构；在烤包子时糖有产生焦化的作用，使产品的色泽和香味更佳。

香油是芝麻中提炼出来的，使包子产生特殊的香味，增加人的食欲。

（二）包子皮面的制作（以发酵面团为例）

原料：特级面粉500g、酵母5g或老酵面50g、白糖25g、大油25g、小苏打4g、食盐4g、水250g。

调制：将老酵面、清水放入盆内，调散，再加入面粉和匀，反复揉搓至面团表面光洁、不粘手、不粘盆时，将面团用湿纱布盖上，静置饧发（春夏季2～3h，秋冬季5～6h）。待面团发酵膨胀，抓一把查看，起蜂窝眼即可，将其倒在案板上，撒上少许扑面、干面粉，再加入小苏打（最好用水溶化）和白糖、融化大油，然后反复揉匀，再静置约10min，即成发酵面团。

包子皮的制作还可添加一些具有特殊口味、营养或功能的原料，如可可粉、蜂蜜、茶叶粉、杏仁浆、花生粉、啤酒、莲子粉等。用添加了上述原料的面团制作的包子食用时唇齿留香，回味悠长，有的还有一定的保健功效。

即便用单一酵面为皮制作的包子，也因酵面的种类不同而风格各异，如用老酵面制作的包子松软、肥嫩、饱满；嫩酵面较韧，组织较紧密，并具可塑

性，成品形态膨胀不过大，不易走形，适宜制作汤包和多种花色包子，如镇江名点蟹黄汤包、苹果包、刺猬包、金鱼包等。

（三）普通包子的制作

普通包子制作工艺流程

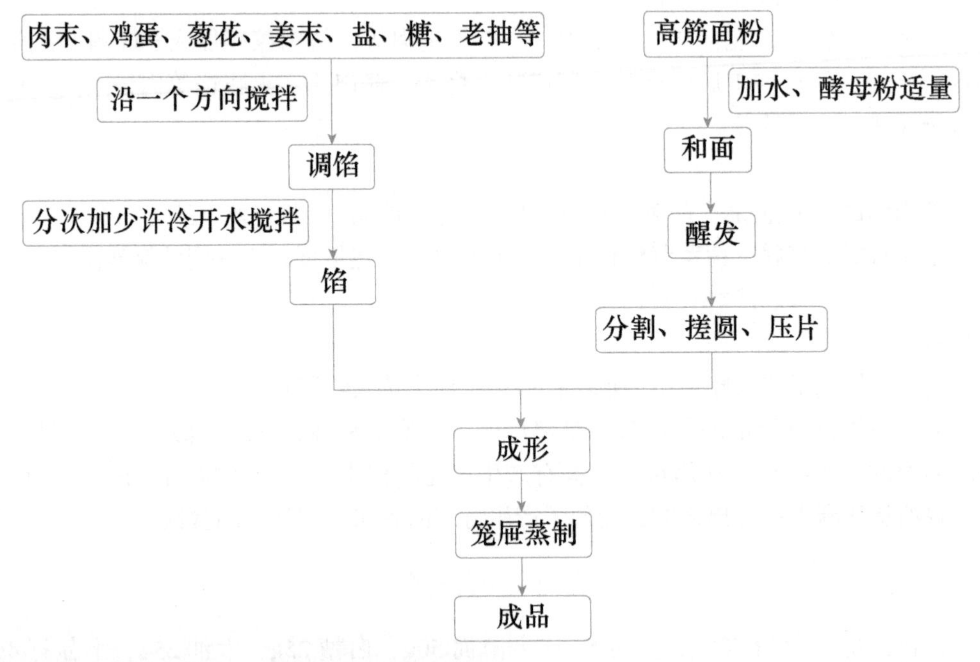

参考配方

包子面皮：面粉250g、水120g、酵母粉3g、大油13g、小苏打2g、食盐3g。

包子馅：羊肉200g、洋葱200g、食盐12g、酱油10g、五香粉8g、胡椒粉5g、味精3g、孜然粉2g、花椒粉3g、鲜姜末5g、味精2g、香油5ml。

操作要点

1. 和面：将原辅料放入容器内混合均匀，搅拌成块，用手揉搓成面团，放在台面上反复揉搓，直至面团光洁润滑，用湿布把面团盖上，醒发30～60min。

2. 制馅：将肥瘦相间的肉剁成肉末，沿一个方向搅拌均匀，在搅肉过程中要加适量的生姜水。同时，酱油要一点一点慢慢加入，以使酱油完全渗到肉里，加完酱油稍等 5min，缓慢加入凉开水，边加边搅拌，之后放入花椒粉、五香粉、食盐、鲜姜末、味精、香油搅拌均匀备用。

3．包馅成形：把发好的面团分割成均匀的小面团，将小面团压扁，擀成直径为 4～6cm 的包子皮，然后加馅包好。

4．熟制：将成形的包子置于笼屉中，隔水蒸 15min 即可。

（四）灌汤包子的制作

灌汤包子制作工艺流程

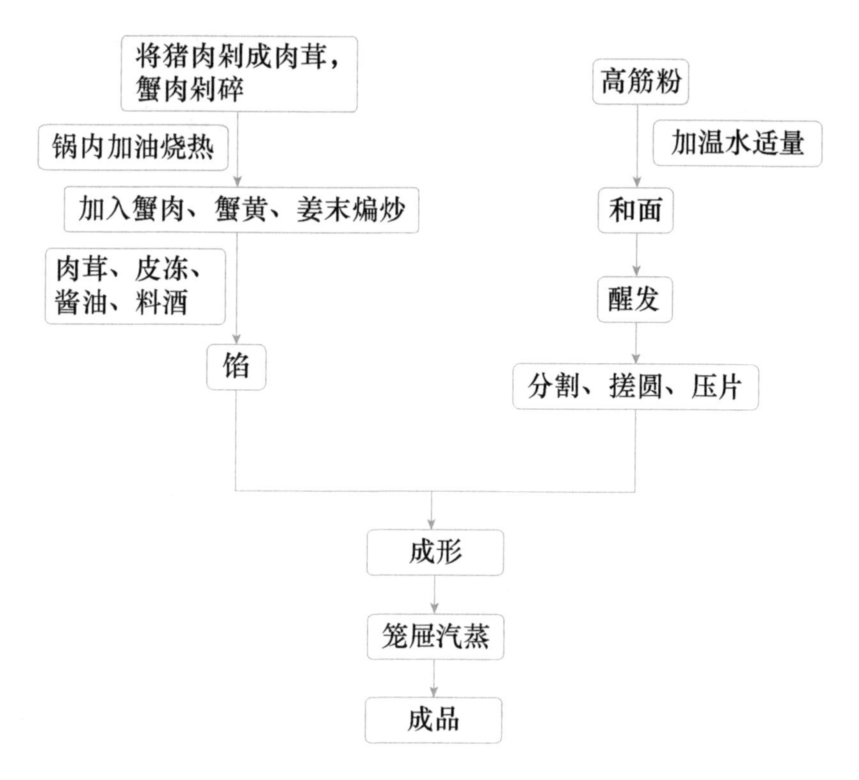

参考配方

包子面皮：高筋粉 500g、温水 300g、酵母粉 3g、小苏打 3g、食盐 3g。

包子馅：猪五花肉 350g、肉皮冻 150g、蟹肉 80g、蟹黄 40g、酱油 20g、猪油 50g、料酒 5g、香油 5g、白糖 5g、葱花 5g、姜末 5g、食盐 10g、胡椒粉 2g、味精 2g。

操作要点

1．面团制作：按发酵面团进行制作。

2．馅的制作：将猪肉剁成肉茸，蟹肉剁碎，锅内加猪油烧热，放入蟹肉、蟹黄、姜末煸出蟹油，与肉茸、皮冻、酱油、料酒等调拌成馅。

3．包馅成形：将面团搓成长条，揪成面坯（每个 50g），擀成圆皮，加馅捏成提褶包。

4．熟制：上蒸笼用旺火蒸 10min 即可。

（五）生煎包子的制作

生煎包子制作工艺流程

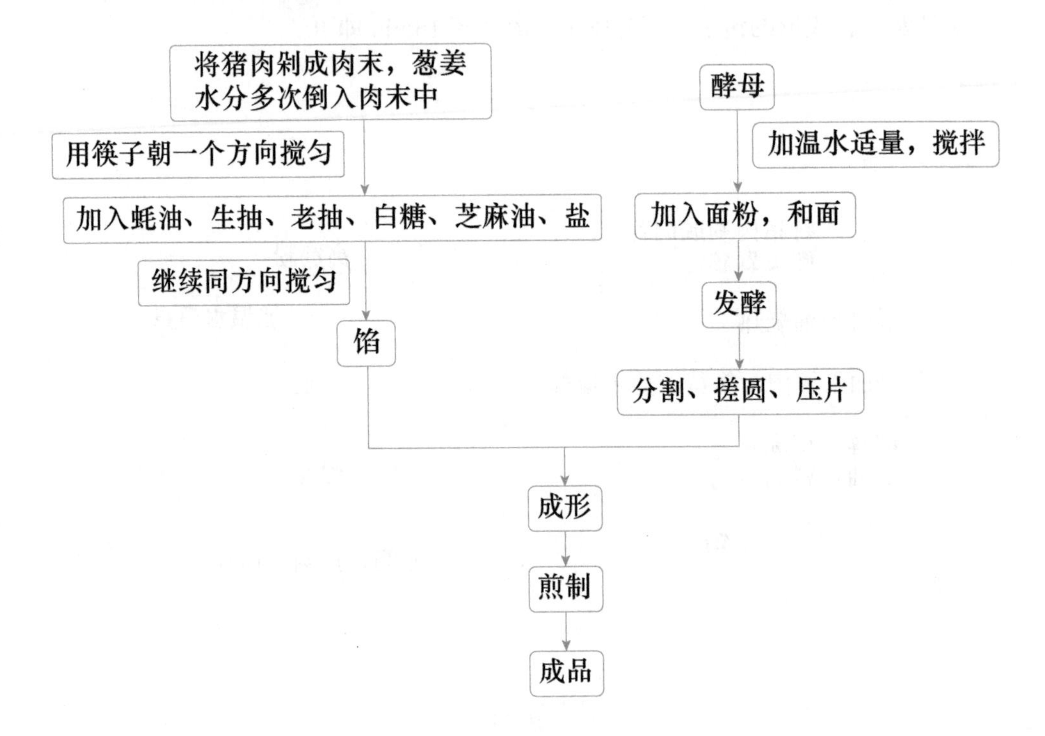

参考配方

包子面皮：普通面粉 700g、高筋面粉 300g、温水 600g、酵母 12g、食用油（菜籽油）200g。

肉馅：猪五花肉 400g、瘦肉 200g、蚝油 10g、生抽 5g、老抽 5g、芝麻 30g、酱油 20g、白糖 20g、芝麻油 10g、十三香 10g、葱 10g、生姜 10g、食盐 10g、味精 5g。

操作要点

1．和面：将酵母放入温水中，拌匀后稍稍静置一会儿使其完全溶解，将酵母溶液倒入面粉中，一边倒一边用筷子搅拌，搅匀成棉絮状，用手将其抓揉成团，反复揉压，揉压至其变成表面光滑的面团，将揉好的面团放入盆中，盖上盖子或湿纱布，于 30℃ 静置 1.5～2h，待其发酵。

2．制馅：在面团发酵的时候准备馅料，将猪肉剁成肉末；葱姜切丝，放入开水中浸泡一会儿，过滤出葱姜弃之，将葱姜水分多次倒入肉末中，一边倒一边用筷子朝一个方向搅匀，再加入蚝油、生抽、老抽、白糖、芝麻油、盐，继续朝一个方向拌匀，待用。

3．包馅成形：面团发酵至2倍大，面团里面充满蜂窝状的小孔时，说明已经发酵好了。将发酵好的面团取出放在撒有一层薄粉的案板上，充分地揉压出里面的空气，要将其再次揉成表面光滑的面团，再盖上湿纱布，静置松弛约10min。将面团切成两份，再分别揉搓成长条，用刀切开成面团，50g面团分成4个剂子，搓圆、压扁，然后擀成中间略厚边缘稍薄的面皮，取适量的馅料放入面皮中央，捏出褶子，收口成包子。

4．煎制：平底锅烧热，放入食用油200g，将包子摆放其中，小火煎约1min；沿锅边淋水，约淹至包子的1/3处，然后盖上锅盖，转中火，焖煮2min；第二次淋水，盖上锅盖，焖至锅内水干，再淋入少许油，将包子煎至底部金黄酥脆，最后撒上芝麻和葱花即可。也可将包子制作好后，先放置一段时间后上笼蒸5min；蒸好后，放入锅中倒入少许的食用油煎，锅要旋转使包子不会粘在锅上，煎成金黄色撒上芝麻和葱花即可。

五　包子的风味特色

包子选料精良，皮薄馅大，表面光亮细腻，花纹整齐，形态多样，鲜嫩适口，肥而不腻，口味醇香鲜美。

包子是发酵食品，它既是主食，又兼副食；既有肉类，又有蔬菜，含有多种营养素，符合平衡膳食的要求。

参考文献

冯明会，陈援援，罗文．2019．包子多汁鲜肉馅心的工艺研究［J］．四川旅游学院学报，(04)：21-25．
唐雄壮．2008．家庭烤包子生产工艺［J］．农村新技术，(03)：37．
王冰洁．2019．各地包子，营养有长短［J］．工友，(04)：58-59．
王静．2013．唇齿留香烤包子［J］．农产品加工，(01)：17．
尹贺伟．2015．浅谈中式面点的开发与创新——以包子制作为例［J］．职业，(05)：140．
张若屏．2016．南瓜包子加工工艺及品质的研究［J］．科技视界，(23)：175，210．
赵敏．2011．小笼包风味特点及形成因素［J］．民营科技，(12)：149．
周中凯，杨雪．2017．分析包子在不同加工与储藏条件下的菌群多样性［J］．粮食与油脂，30(09)：39-42．
朱在勤，陈霞，毛羽扬，等．2006．扬州包子的工艺标准化研究［J］．农产品加工（学刊），(12)：8-12．
朱在勤，崔慧．2014．扬州包子蒸制工艺研究［J］．扬州大学烹饪学报，31(02)：14-18．

饺　子

Jiaozi/Chinese dumplings

Jiaozi is a kind of traditional flour-made food with a long history in China, which is often eaten to celebrate Chinese New Year as its pronunciation is similar to the meaning of "transition from old year to new year" in Chinese. It expresses people's yearning and appeal for a better life.

Jiaozi originated from the Eastern Han Dynasty (AD25-220) , and it is said that Zhang Zhong Jing, the Chinese Medicine Sage, invented it for medical purpose. He made jiaozi stuffed with food that have warming function (e.g. mutton and pepper) to dispel coldness and dampness of a patient. Besides, eating such jiaozi could prevent frostbite on ears during winter time. Traditionally, people would eat jiaozi at Chinese New Year Eve to get rid of bad luck in the old year, or in occasions when people go back home after traveling afar for celebrating the reunion.

In the long process of its development, jiaozi used to have many other names. In ancient times, it used to be called "laowan" "bianshi" "jiaoer" and "fengjiao" etc. During the Three Kingdoms period (AD220-280) , it was called "crescent wonton". In the Southern and Northern

Dynasties (AD420-589) , it was called "wanton". In the Tang Dynasty (AD618-907) , it was called "moon-shaped wonton". In the Song Dynasty (AD960-1279) , it was called "jiaozi" (similar pronunciation but different characters in Chinese) . In the Yuan Dynasty (AD1271-1368) , it was called "bianshi". In the Qing Dynasty (AD1636-1912) , it was finally called "jiaozi", and this name has been used all the time since then.

Jiaozi banquet was originated from the Tang Dynasty, dating back to more than a thousand years ago. At that time, Chang'an (Xi'an city) was the capital city and also the economic and cultural center of the Tang Dynasty. In Chang'an city, a high-class banquet called "Burning Tail Banquet" was very popular, which was a sumptuous feast prepared by the court's official for the emperor after he was promoted. In a food list preserved from the Tang Dynasty for the banquet, it was found that there was a dish called "24 solar terms wonton", which was jiaozi of different shapes and stuffings based on characteristics of each solar term.

The jiaozi banquet was originally comprised of 108 varieties of jiaozi, and the number already exceeded 108 today. In order to meet the needs of different guests in Xi'an, the jiaozi banquet was divided into a welcome banquet, a royal court banquet, a lucky banquet, and a dragon and phoenix banquet, and a variety of skills such as sculpture, molding, embellishment were applied to make jiaozi for these banquets. Thousands of people visit Xi'an for the jiaozi banquet every day from all corners of China. Jiaozi has become a symbol of Chinese culture and tradition, and it is a unique way to treat guests from all over the world to Xi'an.

The difference of jiaozi is mainly reflected in the difference of its stuffing, and each stuffing has its own meaning. For example, the pronunciation of Chinese cabbage sounds like hundreds of wealth (treasures) , so jiaozi with Chinese cabbage stuffing is a blessing for wealth, and it also means everlasting love for the newlyweds. The pronunciation of celery sounds like diligence or hardworking in Chinese, so jiaozi with celery stuffing is a blessing for wealth and diligence.

Jiaozi is characterized by thin flour wrapper and tender stuffing, delicious taste and unique shape. Jiaozi can not only satisfy consumers' taste, but also has rich nutrition. From the point of view of nutrition, jiaozi, as a traditional food in China, will develop faster and be healthier if it is in line with some studies on modern food nutrition. Methods in these studies can help improve the content of vitamins and trace elements, reduce fat and promote protein structure in jiaozi.

饺子是一种历史悠久的民间面食，为传统美食，是中华美食的代表，表达着人们对美好生活的向往与诉求。因取“更岁交子”之意，所以深受老百姓的欢迎。

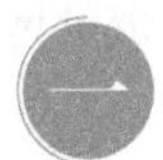

一 饺子的历史起源及发展

饺子又名娇耳、扁食、煮角、籦扎、子孙饽饽，古时有“牢丸”“扁食”“饺饵”“粉角”等名称，是新旧交替之意。远方的人们都会跋山涉水回乡和家人过冬节吃饺子，以示有个圆满的归宿。

饺子起源于东汉时期，为医圣张仲景首创。当时饺子是药用，张仲景用面皮包上一些祛寒的药材（羊肉、胡椒等）用来治病，避免患者耳朵上生冻疮。

饺子在三国时期称“月牙馄饨”，南北朝称“馄饨”，唐代称“偃月形馄饨”，宋代称“角子”，元代称“扁食”，清朝则称为“饺子”。饺子一般要在年三十晚上子时以前包好，待到半夜子时吃，这时正是农历正月初一的伊始，吃饺子取“更岁交子”之意。“子”为“子时”，“交”与“饺”谐音，有“喜庆团圆”和“吉祥如意”的意思。

从20世纪80年代开始，随着我国经济的发展，生活节奏加快，市场上出现了速冻水饺。速冻水饺是将包好的饺子经过速冻以达到冷藏，可以随时食用的一种食物。速冻水饺的产生，使水饺从手工制作转变为工业化生产。

二 饺子的逸闻趣事

（一）饺子的赞美诗

清代何耳的《水饺》诗：

略同汤饼赛新年，荠菜中含著齿鲜。
最是上春三五日，盘餐到处定居先。

（二）饺子宴

用饺子做宴席起源于唐代。当时长安城里盛行一种高等级的宴席，叫作“烧尾宴”，是朝廷大臣官位提升后进献给皇帝的丰盛的大餐。现保存的唐代“烧尾宴”食单里面，就发现有一道菜肴叫作“二十四节气馄饨”，根据二十四个节气，来包成不同形状、不同陷的饺子。

饺子宴最初由108种饺子组成，现如今它的数量早已突破了108这个数字。西安饺子宴分为迎宾宴、宫廷宴、吉祥宴、龙凤宴。饺子的造型综合雕塑、捏塑、点缀等多种技艺，其花色以中国的民间传说或历史典故为依据。

The most popular fruit in the world

饺子宴中的压轴菜“珍珠火锅饺子”相传与慈禧太后有关。据传，当年慈禧太后与光绪皇帝避难西安。一晚慈禧看完秦腔后感到腹饿，即命御厨赶做她从来没有吃过的夜宵。厨师深知太后的癖好，就挖空心思，用鸡鸭鲜菇汤料，做成小拇指大小的珍珠饺子，用双龙紫铜火锅烹制。在夜幕下，燃烧的火焰不断跳动，恰似盛开的朵朵菊花争奇斗艳，异彩纷呈。因此，这道菜又被称为“太后菊花火锅”。

（三）饺子馅的寓意

饺子的不同主要体现在饺子馅的各异，其寓意也各不相同。如芹菜馅，寓意勤财，故为勤财饺；韭菜馅，寓意久财，故为久财饺；白菜馅，寓意百财，故为百财饺；香菇馅，寓意鼓财，故为鼓财饺；酸菜馅，寓意算财，故为算财饺；油菜馅，寓意有财，故为有财饺；鱼肉馅，寓意余财，故为余财饺；羊肉馅，寓意洋财，故为洋财饺；牛肉馅，寓意牛财，故为牛财饺；大枣馅，寓意招财，故为招财饺；甜馅，寓意添财，故为添财饺；野菜馅，寓意野财，故为野财饺。另外，在包饺子时人们常常将糖、花生、枣和栗子等包进馅里。吃到糖的人，来年的日子更甜美，吃到花生的人将健康长寿，吃到枣和栗子的人将早生贵子。

 饺子的分类

饺子主要按地域和国家进行分类。

（一）按地域分类

中国各地饺子的名品甚多，如广东用澄粉做的虾饺、上海的锅贴饺、扬州的蟹黄蒸饺、山东的高汤小饺、沈阳的老边饺子、四川的钟水饺等，都是深受人们喜爱的品种。

（二）按国家分类

按国家不同可分为朝鲜饺子（馅为牛肉、辣椒）、越南饺子（馅为鱼肉、橙皮、猪肉、鸡蛋）、俄罗斯饺子（馅为牛肉、胡萝卜、鸡蛋、葱头和辣椒末）、印度饺子（用料、做法与俄罗斯饺子相似，烤制）、墨西哥饺子（馅为洋葱、牛肉、番茄、荷兰芹菜，饺子皮用手压成长方形。包好的饺子放入用番茄、辣椒、洋葱煮好的调味汤里煮，吃完饺子再喝汤，“原汤化原食”）、意大利饺子（馅为干酪、洋葱、蛋黄及菠菜、牛肉、鸡肉、干酪，主要调料有黄油、洋葱、柠檬皮、肉豆蔻）、匈牙利饺子（以果酱和腌制的李子、杏、乌梅做馅）、日本饺子（馅常采用章鱼和香姜混合物，有一股浓浓的海鲜味和姜的鲜味，常将饺子煎着吃，并配上鱼汤一起食用）、哈萨克斯坦饺子（馅为羊羔肉或者牛肉，加上香辛料和黑胡椒，是一种蒸饺。蒸好后在饺子上淋上黄油、酸奶油或者洋葱酱食用）。

四　饺子的制作工艺

（一）原辅料介绍

饺子之主料——面粉

面粉选用优质、洁白、面筋度较高的特制精白粉或特制水饺专用粉。潮解、结块、霉烂、变质、包装破损的面粉不能使用。由于新面粉中存有蛋白酶的强力活化剂硫氢基化合物，往往影响面团的拌合质量，从而影响饺子制品的质量。因此若用新面粉制作饺子，可在新面粉中加一些陈面粉。

饺子之主料——原料肉

饺子的原料肉选用经兽医卫生检验合格的新鲜肉或冷冻肉。严禁冷冻肉经反复冻融后使用，因为这样不仅降低了肉的营养价值，而且也影响肉的持水性和风味，使水饺的品质受影响。冷冻肉的解冻程度要控制适度，一般在20℃左右的室温下解冻10h，中心温度控制在2～4℃。原料肉在清洗前剔骨去皮，剔除淋巴结及严重充血、淤血处和色泽气味不正常的部分，对肥膘还应修净毛根等。修好的瘦肉肥膘用流动水洗净沥水，绞成颗粒状备用。

饺子之辅料——蔬菜

蔬菜要鲜嫩，除尽枯叶、腐烂部分及根部，用流动水洗净后在沸水中浸烫，要求蔬菜受热均匀，浸烫适度，不能过熟。浸烫之后迅速用冷水使蔬菜在短时间内降至室温，沥水绞成颗粒状并挤干菜水备用。烫菜数量应视饺子数量而定，不可多烫，放置时间过长使烫过的菜“回生”或用不完冻后再解冻使用都会影响饺子的品质。

饺子之其他辅料

饺子的其他辅料，如糖、盐、味精等辅料应使用高质量的产品。对葱、蒜、生姜等辅料应除去不可食用部分，用流水洗净，斩碎备用。

（二）饺子的制作

饺子制作工艺流程

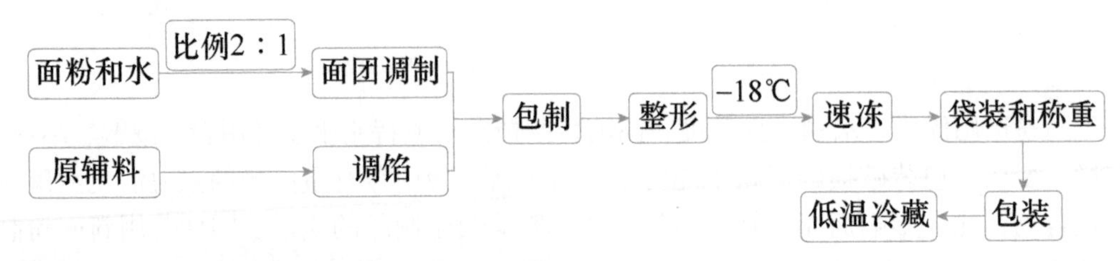

饺子馅参考配方

饺子因馅料不同而异，馅料决定饺子的种类。几种常见饺子馅料配方如下。

猪肉莲白馅：猪肉500g、卷心菜1000g、姜末15g、葱花30g、食盐15g、胡椒粉5g、料酒25g、味精15g、香油25g、精炼油25g。

羊肉馅：净羊肉500g、韭黄250g、姜末50g、葱花50g、花椒5g、鸡蛋2个（约150g）、食盐5g、胡椒粉3g、料酒15g、酱油20g、香油25g、花生油25g。

牛肉馅：牛肉500g、白萝卜1000g、洋葱50g、鸡蛋一个（约75g）、姜末50g、嫩肉粉5g、食盐10g、胡椒粉5g、料酒15g、酱油25g、味精15g、香油25g、精炼油30g、淀粉50g。

三鲜馅：鲜虾仁200g、水发海参100g、冬笋150g、猪前胛肉200g、姜片10g、葱节20g、姜末20g、葱花50g、鸡蛋1个（约75g，蛋清留用）、食盐10g、胡椒粉3g、料酒30g、味精10g、鸡精10g、白糖10g、香油25g、高汤350g。

鸡肉冬笋馅：鸡脯肉750g、冬笋100g、葱花50g、香油5g、姜末20g、食盐10g、味精10g、高汤150g。

鱼肉韭黄馅：去皮鱼肉700g、肥膘肉50g、韭黄200g、葱花50g、料酒20g、姜末5g、食盐15g、味精5g、高汤200g。

茴香馅：茴香500g、猪肉馅375g、料酒5g、酱油10g、盐5g、芝麻油20g。

操作要点

1. 面团调制：面粉在拌合时一定要做

到计量准确，加水定量，适度拌合。要根据季节和面粉质量控制加水量和拌合时间，气温低时可多加一些水，将面团调制得稍软一些；气温高时可少加水甚至加 4℃左右的冷水，将面团调制得稍硬一些，这样有利于水饺成形。

特别要注意和面时间和加水量。和面时间小于 5min 时，面粉不能充分吸收水分，很难形成面团；和面时间超过 15min 时，面筋网络被破坏，导致面团变黏、变软，最佳和面时间为 10min。加水量少面团不能充分形成，面团表面可见干面粉；加水量过大面团太软，变黏；面粉和水的比例约为 2∶1 时，和成的面团表面光滑，无黏性，延展性和弹性适中。

调制好的面团用洁净湿布盖好，防止面团表面风干结皮，饧 5min 左右，使面团中的粉粒充分吸水，更好地生成面盘网络，提高面团的弹性和滋润性，使制成品更爽口。面团的调制技术是成品质量优劣和生产操作的关键因素。和面时加入小麦淀粉和玉米淀粉，包出的饺子晶莹剔透，鲜香四溢，色香味俱全。

2．饺子皮的制作：皮的做法通常有擀和捏两种方式。

第一种方法：擀，把饧好的面团放在案板上，搓成直径 2～3cm 的圆柱形长条。把柱条揪（或切）成长约 1.5cm 左右的小段，用手压扁。再用擀面杖擀成直径 4～7cm、厚 0.5～1mm、中心部分稍厚些的饺子皮。擀皮时，案板上要撒些干面（浮面），以防粘到板上。

第二种方法：捏，先将面剂子揉成扁圆形，然后一边用双手手指捏压，一边旋转。捏成后，皮呈碗状（而擀的皮呈平面状），且所带干面较少，所以更易包。缺点是捏皮比擀皮耗时多。

3．饺馅配制：饺馅配料要考究，计量要准确，搅拌要均匀。要根据原料的质量、肥瘦比、环境温度控制好饺馅的加水量。通常肉的肥瘦比控制在 2∶8 或 3∶7 较为适宜。在高温夏季还必须加入一些 2℃左右的冷水拌馅，以降低饺馅温度，防止其腐败变质并提高其持水性。向馅中加水必须在加入调味品之后（即先加盐、味精、生姜等），否则，调料不易渗透入味，而且搅拌时搅不黏，水分吸收不进去，制成的馅不鲜嫩。加水后搅拌必须充分才能使绞馅均匀、黏稠，制成水饺制品才饱满充实。如果搅拌不充分，馅汁易分离，水饺成形时易出现包合不严、烂角、裂口、汁液流出现象，使水饺煮熟后出现走油、漏馅、穿底等不良现象。如果是菜肉馅水饺，在肉馅基础上再加入经开水烫过并绞碎、挤干水分的蔬菜一起拌和均匀即可。

4．水饺包制：水饺包制是水饺生产中极其重要的一道技术环节，它直接关系到水饺形状、大小、重量、皮的厚薄、皮馅的比例等质量问题。

花边饺子：取饺子皮放上饺子馅，将饺子皮折起来捏紧中间部分；左右两边向里捏紧，再次将边捏紧，然后卷花边。

荷包饺子：取饺子皮放上饺子馅，将饺子皮对折捏紧；从右至左卷花边。

四角饺子：取饺子皮放上饺子馅，将中间捏紧；将左右两边取中间部分捏紧，将其余的边捏紧。

三角饺子：取饺子皮放上饺子馅，先将左边的部分对折捏紧，然后将右边的中间部分向中间捏紧，剩下的两个边捏紧。

月牙饺子：取饺子皮放上饺子馅，将中间捏紧；用手将右半部分的上皮做一个小波浪捏紧，再继续做两个小波浪，左半部分也以相同的方式捏紧收边即可。

5．整形：包制后的饺子，要轻拿轻放，手工整形以保持饺子良好的形状。在整形时要剔除一些如瘪肚、缺角、开裂、异形等不合格饺子。如果在整形时，用力过猛或手拿方式不合理，排列过紧相互挤压等都会使成形良好的饺子发扁，变形不饱满，甚至出现汁液流出、粘连、饺皮裂口等现象。整形好的饺子要及时送至速冻间进行冻结。

6．水饺在速冻间中心温度达－18℃即完成速冻。

7．装袋称重包装。

（1）装袋：速冻水饺冻结好即可装袋。在装袋时要剔除烂头、破损、裂口的饺子以及连结在一起的两连饺、三连饺及多连饺等，还应剔除异形、落地、已解冻及受污染的饺子。不得装入面团、面块和多量的面粉。严禁包装未速冻良好的饺子。

（2）称重：要求计量准确，严禁净含量低于国家计量标准和法规要求，在工作中要经常校正计量器具。

（3）排气封口包装：包装袋封口要严实、牢固、平整、美观，生产日期、保质期打印要准确、明晰。装箱动作要轻，打包要整齐，胶带要封严粘牢，内容物要与外包装箱标志、品名、生产日期、数量等相符。包装完毕及时送入低温库。

五 饺子的风味特色

饺子的特点是皮薄馅嫩，味道鲜美，形状独特。通常以馅料味道好，口感佳，颗粒感好有滋味，紧实成团，颜色鲜亮者为佳。

饺子从食品营养的角度来看，如果和现代食品营养学结合起来，开发更多的品种，提高维生素和微量元素含量、减少脂肪含量、改善蛋白质结构等，将会使饺子更健康。

参考文献

陈洁，霍蓓，宋泽伟，等．2013．小麦粉粉质特性与饺子皮质构品质的关系研究［J］．食品工业，34（04）：136-138．

程云．2011．关注产品：速冻水饺［J］．标准生活，（9）：78-79．

黄忠民，王笑，潘治利，等．2018．速冻饺子馅料中蔬菜预处理方式工艺优化［J］．农产品加工，（18）：21-25．

兰静，傅宾孝，Assefaw E，等．2010．饺子的实验室制作与品质评价方法研究［J］．食品科学，31（03）：136-141．

李娟，葛斌权，陈正行，等. 2019. 马铃薯生全粉对饺子皮蒸煮特性和质构特性的影响［J］. 中国粮油学报，34（09）：28-32.

林莹，辛志平，古碧，等. 2011. 两种羟丙基淀粉对速冻饺子品质影响［J］. 粮食与油脂，（08）：10-13.

秦兰田. 1992. 从速冻饺子“热”谈中国式快餐食品［J］. 商业研究，（03）：49-50.

谭亦成，陈丽，张喻. 2013. 饺子皮中常用辅料的研究进展［J］. 农产品加工（学刊），（22）：53-56.

王岸娜，李秀玲，吴立根. 2012. 玉米种皮膳食纤维对面团流变学特性及饺子皮品质的影响［J］. 河南工业大学学报（自然科学版），33（04）：5-10，16.

王向明. 2002. 浅论速冻水饺现状及发展趋势［J］. 食品科技，（12）：40-41.

张中义，柴颖，范雯，等. 2018. 大豆分离蛋白对速冻饺子肉馅抗冻性能的改善［J］. 食品工业，39（01）：30-34.

金华酥饼

Jinhua Crisp Cake

Jinhua crisp cake, also known as dried crisp cake, is a traditional cake of Jinhua city in Zhejiang province. It is golden in color and tastes crisp and delicious. Its crust is as thin as the paper and its stuffing tastes moist but not greasy after baking. The crisp cake has an unique attraction with its strong fragrance and fresh salty taste. It had been well known in the Ming Dynasty, and there was a sentence passed on with Li Bai's approval that "the aroma of crisp cake attracted me down from the horse".

Jinhua crisp cake enjoys a long history. The method of making pulp cake was recorded in *Qi Min Yao Shu* wrote by Jia Sixie in Northern Wei Dynasty and this can be regarded as the early form of crisp cake. Jinhua crisp cake was firstly recorded in the first Chinese recipe collection *Zhong Kui Lu* written by Wu in the Southern Song Dynasty (Wu came from Wu Zhou, now Pujiang county of Jinhua city). "The Making Method of Crisp Cake", in one of its sections about making cakes, had a detailed record about the making of Jinhua crisp cake at that time, which dated back to 800 years ago. According to the statistics, there are eighteen books about the legend and production of Jinhua crisp cake. *Jinhua City Records* compiled the term "Jinhua crisp cake" and made a detailed introduction. In addition to historical records, there is also folklore about Jinhua crisp cake in Jinhua city. The legend has it that the Jinhua crisp cake was created by "the devil king" Cheng Yaojin, who was regarded as the founder of Jinhua cake.

There are many beautiful legends about Jinhua crisp cake telling the stories of famous historical celebrities and it. The most famous one is that "Li Bai got off the horse at once after smelling its fragrance". According to legend, Li Bai, the famous poet in the Tang Dynasty, came to Wuzhou. When he entered the city, he smelled the charming fragrance and wanted to seek the source of the fragrance. Finally he found the cake shop and tasted the delicious crisp cakes. After Zhu Yuanzhang conquered Jinhua area in the Ming Dynasty, he and his military advisor Liu Bowen tasted the cakes in the Mingyue building of Wuzhou while they were discussing political issues. During the period of Taiping Heavenly Kingdom, people of Jinhua city once comforted Li Shi Wang with the cakes who had repeatedly defeated the Qing army.

Crisp cake is not only closely related to celebrities, but also the ordinary people. As a kind of food that can be preserved for a long time, it is more popular among the common people. Crisp cakes are indispensable snacks when people are going out for examination and business. Rural people often use crisp cake as a dessert and they usually eat a crisp cake with water after hard work. On major festivals, children often carry a few bags of crisp cakes back home to show the filial respect for the elders of the family; in the wedding festival, crisp cake is often used as a return gift.

As Jinhua crisp cake has been rarely known to the world, the Jinhua Crisp Cake Industry Association decided to participate in the 31st Panama Expo. In order to make the judges taste the most local Jinhua crisp cake, the Association decided to dispatch the most skilled cake masters. A dozen of the finest crisp cake makers in Jinhua made the best Jinhua crisp cakes in 3 days. The Association picked out the best and sent them to Panama.

How can we display the most distinctive aspect of Jinhua crisp cake to the judges? The Association President Huang Weijian put on a full set of clothes for crisp cake master, and then filled the counter with introduction albums printed in English and Chinese. The Guizhou Ermao wine booth was next to the crisp cake booth. When the judges tasted the wine one by one, Huang Weijian quickly told the interpreter to show the judges that in Jinhua city, it was perfect to sip wine and eat crisp cakes. The judges thumbed up after Huang Weijian teached them how to appreciate the cake.

Jinhua crisp cake is flat and round like crab shell, and golden in color with a slightly lustrous surface, covered with sesame. It is stuffed with preserved vegetables and pork. Both its upper side and lower side have more than 10 layers and each layer is as flat and thin as paper. The filling is oily but not greasy, and the cake tastes salty and delicious.

金华酥饼又名梅干菜酥饼，色泽金黄，香脆可口，是浙江省金华地区的传统名点。其面皮分层薄如纸，烤制后酥松油润而不腻，形如蟹壳，两面金黄，味道极佳，入口酥碎，遇湿消融；还以浓烈的陈香和鲜咸的回味显示其特有的魅力。金华酥饼明代已闻名于世，民间更有李白“闻香下马”的美谈。

一 金华酥饼的起源及发展

金华酥饼历史悠久，北魏贾思勰《齐民要术》中，记录的髓饼制作法，被视为酥饼制作的早期形态：“以髓脂、蜜合和面，厚四五分，广六七寸，便着胡饼炉中，令熟，勿令反复。”“金华酥饼”有史可查最早见诸于南宋时婺州（金华）浦江吴氏所著的中国第一部菜谱《中馈录》中，其中“酥饼方”对当时金华酥饼的制作有详细记载，迄今约有800年历史。据查全国有18本古籍记载“金华酥饼”的传说和制作。《金华市志》收编了“金华酥饼”的条目并做了详细介绍。

除历代文献记载外，金华民间还流传着关于金华酥饼的传说。坊间传说金华酥饼的首创者是“混世魔王”程咬金，至今该行业中仍奉其为祖师。位列唐初凌烟阁二十四功臣之一的程咬金隋末落难到金华以卖烧饼为生。金华人爱吃梅干菜扣肉，他就以梅干菜和猪肉为馅。有一次，烧饼做得太多了，一整天也没卖完，他便将其保存起来，准备第二天继续卖。为使烧饼保质，程咬金将烧饼放在炉边上烘烤过夜。第二天起床发现，烧饼里的油都给烤出来了，饼皮更加油润酥脆，成了酥饼，上市后大受欢迎。随后程咬金将烧饼加以改进，制出的酥饼圆若茶杯口，形似蟹壳，面带芝麻，两面金黄，加上干菜肉馅之香，更具特殊风味。其后程咬金参加了隋末农民起义，成为唐朝开国元勋。他在功成名就之后，仍忘不了早年的卖饼生涯，极力推荐该小吃。“金华酥饼”随着首创者的名气而名扬四海，这种特殊的做法代代相传，名气越来越大，成为闻名遐迩的传统特产。

金华酥饼的发展，大致可分为四个时期。

清末民初，金华酥饼已遍布城乡，较大的村镇均有制作。由于连年战争，百姓民不聊生，从事金华酥饼生产的大多单打独斗，一炉一板一人，手工烤制，现做现卖。而且很多门店都实行多种生产，每个品种量不大，聊以为生而已。金华酥饼传统制作技艺分布以金华婺城、兰溪、浦江为主。

新中国成立初期，金华地区与全国的形势一样，开展了轰轰烈烈的农业合作化。各种小吃门店纷纷取消，其中金华酥饼因其历史悠久，覆盖面广，群众基础强而得以保留，但生产规模主要集中在几家国营企业。

十一届三中全会后，金华酥饼生产迎来了新的发展机遇。以师带徒，以亲带友的方式慢慢地使金华酥饼的从业人员越来越多。加上市场准入门槛低，个体酥饼店如雨后春笋般地涌现，遍布金华市城乡。据不完全统计，金华市有近100多家酥饼店，但大多是小作坊的形式，前店后作坊，年产值较小。

2005年8月，金华酥饼行业协会成立，它是金华酥饼发展史上的一个重要转折点。2013年，金华酥饼历经8年时间，产值从5000万元增加到了5.7亿元。同年举行的巴拿马博览会上，金华酥饼被授予了金奖。

金华酥饼的逸闻趣事

（一）金华酥饼的故事

金华酥饼有许多美丽的传说，讲述了历史名人与金华酥饼的故事，其中最著名的是“李太白闻香下马”。相传唐代诗人李白来婺州，入城即闻到缕缕诱人的香味，引得他下马循香探源，终于寻得饼店，品尝了奇香可口的酥饼。明代朱元璋攻克金华后，曾与军师刘伯温在婺州明月楼一边品尝酥饼一边议政。太平天国时，金华民众曾以酥饼慰劳屡败清军的李侍王。

酥饼不仅与名人有着千丝万缕的联系，与普通老百姓的关系更为密切。作为一种可长期保存的食品，在老百姓中更受欢迎，出门赶考、经商，酥饼是不可缺少的携带糕点。农村人常常将酥饼作为点心，劳苦耕作后吃个酥饼来口水，继续接着干活。逢年过节孝敬家中长辈，儿女们经常拎着几袋酥饼回家看望老人。在结婚的喜庆日子，酥饼也经常作为回礼。

（二）第31届巴拿马国际贸易博览会金奖背后的故事

在第31届巴拿马博览会召开之际，为了让评委尝到最正宗的金华酥饼，行业协会决定派遣精兵强将。十几位金华最顶尖的酥饼制作人在3天时间里制作出最好的金华酥饼，一共做了几十公斤。协会首先挑选出大小相当、重量都在30g的酥饼，接着分批次挑选一只压碎、品尝，味道地道则代表这批次的酥饼都有资格参赛，最后好中选优、精挑细选了100只酥饼，一共3kg。

因到巴拿马要通过美国转机，怎么让酥饼在半个地球的旅程中保存完整？行业协会花了很多心思。首先将每个酥饼都单独包装起来，然后15个一组是一个大包装。大包装又依次放进特质箱子的8个夹层中，外面还有海绵，坐飞机的时候不敢托运，一直随身带着箱子。经过48h的等待、转机、飞行，到达展会现场打开箱子一看，100只酥饼，每只都保存完好。

在狭小的展示摊位上，金华酥饼行业协会会长黄维健穿上整套的黄色酥饼大师服，把印有中英文的酥饼画册摆满柜台。旁边的展位是贵州二茅酒，评委一个个品尝过来的时候，黄维健就赶紧拉着翻译告诉评委，在金华当地，呷着酒吃着酥饼是绝配。评委按照黄维健的“教程”品尝起酥饼后，一个个都竖起大拇指。金华酥饼最终获得第

31届巴拿马国际贸易博览会金奖。

三 金华酥饼的制作工艺

（一）原辅料介绍

金华酥饼的制作原料主要有梅干菜、猪肉、面粉、饴糖、菜籽油、精盐、芝麻等。

金华酥饼之主料——中筋面粉

面粉按蛋白质含量可分为高筋面粉、中筋面粉、低筋面粉及无筋面粉。中筋面粉的蛋白质含量在8.0%～10.5%，颜色介于高、低粉之间，呈乳白色，体质半松散。金华酥饼的酥皮主要由中筋面粉制作而成，因为其中的蛋白质含量能保证其在经特殊发酵和烤制后做到香脆酥软，柔韧适中，达到金华酥饼的制作要求。

金华酥饼之馅料——梅干菜

梅干菜油光乌黑，香味醇厚，耐贮藏。菜料主要有大叶芥、花叶芥和雪里蕻3个品种。芥菜含有硫代葡萄糖苷、蛋白质和矿物质。加工后的腌菜香味独特，滋味鲜美。晒干后的梅干菜呈酱褐色，有独特的菜干香味，用来作为金华酥饼的馅料更是别具一番风味，惹人垂涎。

金华酥饼之馅料——猪肉

金华酥饼选用猪背脊的上等肥肉。猪肉与梅干菜蒸煮后，混入梅干菜独有的香味。作为金华酥饼的馅料，咬上一口，肉的鲜美与菜的酸香配合酥松的脆皮融入口中，回味无穷。

金华酥饼之配料——饴糖

饴糖用在酥饼的酥皮中可调和面皮发酵中产生的碱味，并为面团发酵提供原料，且能使酥皮的味道更加甜美。

金华酥饼之配料——芝麻

芝麻药食两用，被视为滋补圣品。芝麻撒在金华酥饼表面起点缀及增加香味的作用。

（二）金华酥饼的制作

金华酥饼制作工艺流程

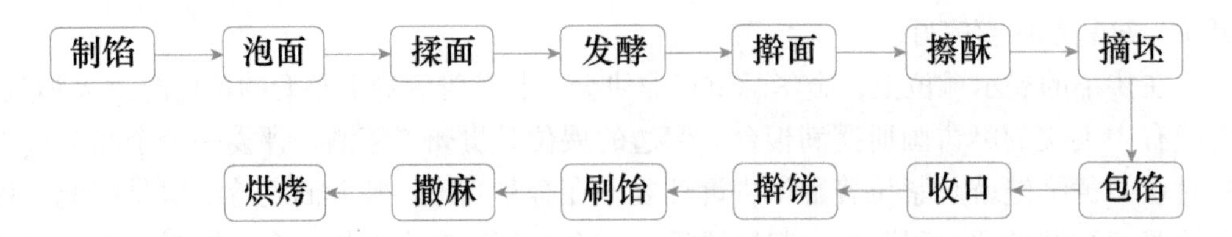

参考配方

水油皮：中筋面粉500g，水250ml，砂糖50g，菜籽油100g，小苏打、碱面适量。

油酥：低筋面粉150g，菜籽油75g。

馅料：半肥半瘦猪肉500g，梅干菜100g，砂糖10g，高度白酒5ml，食盐5g，胡椒粉3g，老姜5g，酱油5ml。

表层：鸡蛋150g（约2个），白芝麻10g，饴糖15g。

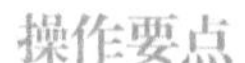

操作要点

1．制馅：先把梅干菜泡水至少1h。把姜和猪肉剁成馅，再加白酒、盐和胡椒粉搅拌均匀，锅里放油，将猪肉馅炒至变色，再加入适量酱油翻炒，最后放入泡好的梅干菜炒均匀。

2．泡面、揉面、发酵：称取适量小麦粉加入热水拌匀，称为泡面，面泡得太熟不起发，泡得太生不松脆，水温为春、冬季90℃，夏、秋季70℃。面团摊开晾凉后加入适量食用小苏打，和成面团揉匀揉透，放置发酵。在适量碱面中加入饮用水配制成碱液，待酵面团具有弹性且呈海绵状时，兑入碱液，反复揉匀揉透，擀成面皮，抹上一层食用植物油，撒上适量面粉用手抹匀，再自外向里卷起，搓成长圆形。

3．擀面、擦酥：擀面的轻重直接影响到酥饼的松脆。150g的面粉和75g的油（最好使用菜籽油）混合成油酥，将面皮擀开成2～3mm厚，然后均匀抹上油酥，将面皮卷起来，边卷边拉长。

4．摘坯：摘坯时要求运用手腕抖动的力量，摘成25g重的面坯，逐个按成中间厚、周边薄的圆皮，使酥层藏在面坯内，称“暗酥”。

5．包馅、收口、擀饼：塞的馅要适中，做的饼不能太扁也不能太鼓（25g面坯约10g馅料），包馅收口要好，层次分明。不偏皮、不空膛、不露馅。收口朝下放在案板上，擀成直径为8cm的圆饼。

6．刷饴：饴糖加入清水调成蜂蜜状，成饴糖水。将饼坯置入烤盘，刷上饴糖水。

7．撒麻：芝麻洗净，放入淘箩内加入沸水，使之松涨，静置片刻后上下翻动，再冲一次沸水，使芝麻粒涨大。在刷过饴糖水的饼上撒上芝麻。

8．烘烤：烘烤时要求掌握好炉温火候，起炉时要熟练快速，才能保证酥饼通体金

黄、完整丰满。烘烤在特制陶炉中进行，内燃木炭，将饼坯贴于炉的内壁，烘烤、焖烘，然后将炉火退净后焙烤，前后在5h以上完成。由于烘烤时间长，饼中水分大多蒸发，利于贮存；即使受潮，烘烤后依然酥香如故。因面皮用面粉搓酥，分层薄如纸，烤制后酥松油润而不腻。

四 金华酥饼的风味特色

金华酥饼扁圆形，大小均匀，形若蟹壳，皮面金黄，色泽均匀，表面略带光泽，上面满布芝麻，中间以干菜肉为馅，上下各有10余层，每层扁薄如纸，馅心油而不腻，香浓味美；酥松油润而不腻，咸香可口。

金华酥饼富含碳水化合物和脂肪，能够为机体提供必要的能量，其配料饴糖、芝麻等均具有一定的营养价值。

参考文献

蒋晓翠，姚波，黄维健. 2019. 脂溶性茶多酚延长金华酥饼货架期作用的研究［J］. 食品工程，（01）：32-35.
金俊，揭良，李泓勇，等. 2016. 棕榈油在金华酥饼皮中的应用研究［J］. 中国油脂，41（03）：70-73.
楼瑞. 2009. 千年传承金华酥饼 工序严谨风味独到［J］. 现代营销（创富信息版），（09）：35.
叶文洁. 2017. 传统食品产业转型升级的挑战及对策——基于对金华酥饼产业的调研［J］. 现代商业，（04）：12-13.
张金焜，孙慧平. 2005. 采用新型烤炉生产金华酥饼的工艺研究［J］. 金华职业技术学院学报，05（02）：1-3.
张金焜. 2009. 传统炭炉烘烤对金华酥饼感官品质的影响分析［J］. 江西农业学报，21（06）：92-93，102.
张金焜. 2009. 金华酥饼智能烤炉的研制［J］. 机电工程，26（07）：108-109.
张金焜. 2009. 酥饼烤制温度的控制模型及仿真分析［J］. 农产品加工（学刊），（08）：76-78.
张金焜. 2012. 基于温度控制器的金华酥饼烤箱电气化改造［J］. 中国农机化，（02）：140-144.

萨其马

Sachima

Sachima, also known as Saqima, Shaqima and Saiqima, a Manchu handmade dessert of Chinese characteristics. Originally, Sachima referred to as "Gounaizi (a kind of Chinese wolfberries) dipped in sugar". With wheat flour, syrup and eggs as its main ingredients, Sachima is made through such steps as making dough, fermenting, laminating, dough recovery, slicing, frying, mixing syrup, molding, decoration and cutting.

Sachima has a history of more than 200 years since its first appearance in the *Yu Zhi Zeng Ding Qing Wen Jian* (*the Revised and Enlarged Manchu Dictionary Compiled to the Imperial Order*) by Fu Huan during emperor Qianlong's rein (1736-1796) in the Qing Dynasty. In Manchu dictionary, Sachima means one kind of Shatangguozi (fried flour wrapped with sugar, also known as Jinsi cake in Chinese) made with sesame and sugar. In the *Revised and Enlarged Dictionary Compiled to the Imperial Order*, the Manchu word for Sachima means it is made by frying dough with sesame oil, dipping it with syrup, and sprinkling clean sesame seeds when translated in Chinese. As there was no proper Chinese word to it, the names were transliterated into Shaqima, Saiqima and so on.

Recorded in *Ma Shen Miao Tang Bing Hang Hang Gui Bei* (a tablet that records rules of sweet pastry industries in Mashen temple) in the 28th year of Daoguang's rein (1848) , Sachima is essential for monks and people of Eight Banners (a way to organize the country and military affairs) , and wedding or funeral. As recorded in *Guang Xu Shun Tian Fu Zhi* (a book that records every aspect in Shuntian Prefecture during Guangxu's Reign (1875-1908)), Sachima was a lama dim sum. Now it is served in the market and tastes much delicious, which was made of fruit and flour and then steamed with sugar and lard. *Annual Customs and Festivals in Peking* (1906) , written during the 32nd year of the reign of Emperor Guangxu, recorded that food in Beijing was linked with the season. After October, there will be Sachima. It was made of rock sugar, cream and wheat flour in the shape of glutinous rice, then cooked in oven with amiantus to form the square shape. It was sweet and delicate to eat. It was a seasonable food in Beijing.

The predecessor of Sachima was "Dagaomudantiaozi" and "Cuotiaobobo". To make Cuotiaobobo, the first step is to beat the steamed rice into dough on the stone repeatedly, and then dip into the soybean flour and pull it into strips. Fry it and cut it into chunks. Up to now, as a delicious pastry of the Manchu ethnic group, Sachima has been brought to all over China from the north along the Jingxi Ancient Road.

A Manchu general whose family name was Sa, served in Guangzhou province during the Qing Dynasty. He loved riding and hunting, and liked to have some pieces of dessert after hunting. He ordered that the dessert should not repeat itself each time. One day, when he went to hunting, he especially told the cook to make something new. If he was not satisfied, he would kill the cook. The cook in charge of the refreshments listened, lost his mind, and accidentally fried the refreshments with the egg mixture. Just at this moment, the general urged the dessert. The cook was so furious that he snapped, "kill the one who rides a horse," and hurried to bring out the refreshments. To his surprise, general Sa was quite satisfied and asked what the name of the snack was. The cook responded, "kill the one who rides a horse", "shaqima" in Chinese pronunciation. As a result, general Sa heard it as Sachima and thus it got its name.

Sachima has the characteristics of beige color and soft and flaky taste. Neither too hard, nor too soft, it is sweet and delicious with rich fragrance of osmanthus honey. Thus, it wins people's love.

萨其马又名沙琪玛、沙其马、赛其马，是中国的特色糕点，满族的一种手工点心，原意是“狗奶子（枸杞子）蘸糖”。萨其马以小麦粉和鸡蛋、糖浆为主要原料，或添加乳粉、芝麻等辅料，经调制面团、静置（发酵）、压片、再醒发、切条、油炸、拌糖浆、成形、装饰、分切一系列工序制成的中式糕点。

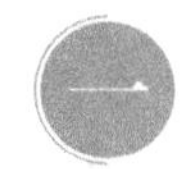

萨其马的起源及发展

萨其马最早见于清朝乾隆年间（1736年～1796年）傅恒等编的《御制增订清文鉴》，至今已有220多年的历史。在满文字典中，萨其马是由胡麻及砂糖制成的一种砂糖果子（汉语称为金丝糕）。在清代满语字典《五体清文鉴》（卷二十七·食物部·饽饽类）、《三合切音清文鉴》和《御制增订清文鉴》中单词对应汉语均为“糖缠”。《御制增订清文鉴》中的满语译为汉语，即白面用芝麻油炸后，拌上糖稀，放洗过的芝麻制成。由于当时找不到汉语代称，便直接将满语音译，所以亦会出现“沙其马”“赛其马”等称呼。

道光二十八年（1848年）的《马神庙糖饼行行规碑》写道“乃旗民僧道所必用。喜筵桌张，凡冠婚丧祭而不可无”。据《光绪顺天府志》记载“赛利马按为喇嘛点心，今市肆为之，用面杂以果品，和糖及猪油蒸成，味极美”。光绪三十二年（1906年）的《燕京岁时记》说：“京师食品亦有关于时令。十月以后，则有……萨齐玛乃满洲饽饽，以冰糖、奶油合白面为之，形如糯米，用不灰木烘炉烤熟，遂成方块，甜腻可食。……京师应时之食品也。”

萨其马前身又叫“打糕穆丹条子”和“搓条饽饽”。制作搓条饽饽先把蒸熟的米饭放在打糕石上用木锤反复打成面团，然后蘸黄豆面搓拉成条状，油炸后切成块，再撒上一层较厚的熟黄豆面即成。时至今日，萨其马作为满族饽饽的美味，已经从北方随京西古道传遍全中国。

萨其马的逸闻趣事

（一）萨其马的传说

清朝，有一位姓萨的满洲将军在广州任职，他擅于射箭，酷爱骑马打猎，每次打猎前后都会吃一点点心，还不能重样。有一次萨将军出门打猎前，特别吩咐厨师要“来点新的玩意儿”，若不能令他满意，就会杀了厨子。当时，厨子正在制作点心，听到此话吓的六神无主，一不小心，把沾上蛋液的点心炸碎了。偏偏这时将军又催要点心，厨子生气大骂一句：“杀那个骑马的！”就满头大汗、惊慌失措地端出点心来。想不到，这点心色泽米黄，口感酥松绵软，香甜可口，香味浓郁。萨将军吃后相当满意，

大加赞赏，并问点心叫什么名字。厨子随口搭话：“杀骑马。”结果萨将军听成了“萨骑马”，传世名点由此而生。

（二）萨其马的爱情故事

拥有悠久历史的美食，总有一段扣人心弦的故事，萨其马也不例外。

清顺治帝福临养女，安郡王岳乐次女的和硕柔嘉公主出生于1652年5月，抚养宫中。出于政治需要，在1658年年仅6岁的公主就被许配给了耿聚忠（靖南王耿仲明之孙，耿继茂第三子），直到和硕柔嘉公主12岁二人才完婚。耿聚忠自小生活在宫里，与顺治、康熙有着深厚的感情，对这门亲事非常满意，对和硕柔嘉公主更是宠爱有加。政治婚姻毕竟不是自己选择的，又加上年纪太小，公主在结婚后婚姻并不是很幸福，于1673年，和硕柔嘉公主去世了。和硕柔嘉公主死后被埋葬在门头沟区龙泉镇，埋葬地被当地人俗称为“公主坟”或“耿王坟”，周边风景秀丽，风水极佳，墓葬规制比同级别的公主坟高。公主的死，给耿聚忠打击十分大，他悲痛欲绝，发誓今生不再继娶。

和硕柔嘉公主生前特别喜食点心，萨其马是其最爱。为此，萨其马成为必不可少的祭祀品。公主去世后第一年祭祀时，作为祭祀品的萨其马被人哄抢而去。耿聚忠得知后决定，每年在祭祀和硕柔嘉公主的时候，都沿途大量发放萨其马。这一行为使当时的北京人都知道有一个“公主坟”在门头沟龙泉镇，而且满族名点萨其马也因耿聚忠的这个举动在北京门头沟民间流传开来。后来，萨其马因其卡路里含量高、耐饥的特点，被常年行走在京西古道的马帮和驼队当作了绝佳的补给美食。

三 萨其马的种类

不同人群对萨其马的口味偏好不同，促使萨其马的种类也越来越多样化，市场上已见的种类有巧克力味、芝麻味、焦糖口味、牛奶口味及台式口味等，研究报道的有绿豆萨其马、玉米萨其马、苦荞萨其马、薏苡仁无铝萨其马、薏苡仁芡实萨其马等。

四 萨其马的制作工艺

（一）原辅料介绍

萨其马的制作原料主要有高筋面粉、鸡蛋（鸡蛋液、全蛋粉）、白砂糖、葡萄糖浆、发酵粉、脱脂奶粉、食盐、小苏打、蜂蜜、食用油（棕榈油、花生油）、枸杞、葡萄干、瓜子仁、芝麻、桂花等。

萨其马之主料——高筋面粉

萨其马的制作需要对面粉进行发酵以及油炸，为保证面条弹性、筋度，多选用高筋面粉。

萨其马之配料——鸡蛋

形成萨其马色泽及综合风味。

萨其马之配料——泡打粉

泡打粉是大多数面制品必不可少的食品添加剂。制作萨其马所用面粉需要进行发酵以达到膨松的效果，泡打粉是影响萨其马膨松效果的重要因素，适量泡打粉可以使萨其马形成膨松多孔的结构，质地柔软而口感酥软。因此，萨其马加工时，需在面粉中掺和泡打粉调制面团。

萨其马之调味——糖浆

萨其马初始时调味以蜂蜜为主，后来多用饴糖作为甜味剂。蜂蜜、细砂糖、麦芽糖是萨其马重要的调剂品。在萨其马加工中起到成形的作用并赋予萨其马绵甜松软的口感，使萨其马入口即化。部分产品将桂花调配于糖浆中，使萨其马具有桂花特有的风味。

萨其马之配料——棕榈油（花生油）

棕榈油富含饱和脂肪酸较多，稳定性较好，不容易发生氧化变质。工业化萨其马生产以棕榈油进行油炸；家庭制作时，可用花生油代替。

萨其马之装饰——枸杞子、葡萄干、芝麻等

过去以“狗奶子”（枸杞子）装饰萨其马，红黄相间，色泽诱人；至乾隆时期，已经不用“狗奶子”，改为撒芝麻。目前，葡萄干、瓜子仁、山楂、枣、青梅等逐渐成为萨其马的果料。各种装饰用原料经洗净，晾干后待用。

（二）萨其马的制作

萨其马制作工艺流程

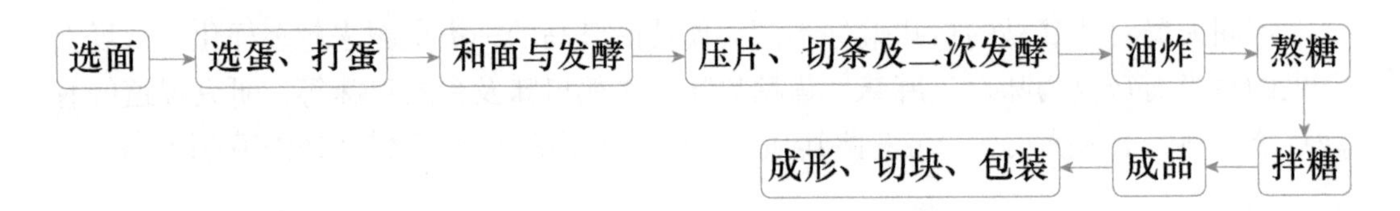

参考配方

配方一：高筋面粉1000g，鸡蛋700g，食用油1000g，白砂糖1000g，饴糖1200g，蜂蜜300g，发酵粉20g，葡萄干150g，瓜子仁50g，麻仁80g，桂花160g，青梅250g，淀粉600g。

配方二：高筋面粉250g、鸡蛋210g、食用油250g、白糖100g、麦芽糖浆150ml、泡打粉7g、葡萄干50g、核桃碎50g、松仁50g、芝麻30g。

操作要点（以配方一为例）

1. 选面：一般要选择新鲜的高筋面粉，不宜选择陈面，更不宜选择虫蛀、鼠咬、霉变的污染面粉。只有新鲜的高筋面粉，蛋白质含量高，才能保证萨其马的制作成功。

2．选蛋、打蛋：新鲜鸡蛋是最基本的要求。配方中增加鸡蛋的量能使萨其马达到更好的口感，其中添加65%鸡蛋的配方品质最优。先在流动清水中将新鲜鸡蛋外壳清洗干净，放在篮中沥干后将鸡蛋液打入洁净的贮桶中。然后边加鸡蛋边搅拌，对萨其马改良效果最好，有利于面团的充气，使煎炸过程中面丝更容易膨大。

3．和面与发酵：将面粉和泡打粉、小苏打、全蛋液等配料搅拌均匀至面筋形成，表面光滑、内部无面结为止。面团和好后置入发酵箱内，进行第一次发酵，30～40℃发酵4h。

4．压片、切条及二次发酵：把醒好的面团倒在洁净工作台上，用手工擀杖或压面机压成宽22cm、厚0.15～0.2cm薄片，然后切成约长20mm、宽5mm、厚1.5mm大小的面条，面块表面应涂抹适量淀粉防粘。将成形面条均匀平铺于塑料醒发盘内于室温下进行二次发酵60～180min。将发酵好的面条筛去扑面，进行油炸。

5．油炸：将食用油倒入锅中，烧至120～160℃，倒入适量面条，炸至淡黄或金黄色，熟透捞出。油炸用油要清亮，注意控制油温，油温过高、炸油使用太久，或油炸时间过长则色泽偏深，油温过低则不膨松。

6．熬糖：砂糖加入适量水放入锅内（糖：水为4：1）用小火加热，直到砂糖溶解，然后大火烧开，加热到115℃左右时加入糖浆和桂花。继续以小火熬煮，直至出现浓密泡沫，稍稍冷却后能扯出细线，准备挂浆（又称套糖）。

7．搅拌：将炸面条与熬制的糖浆混合均匀。膨松度比较好的萨其马条表皮有细密的小孔，口感比较好，并能吸收糖浆。糖浆是靠表皮的气孔渗透到内部去的，如果表皮光滑，糖浆只能挂在表面。

8．成形、切块、包装：首先挑选小料，青梅加工切片，葡萄干、瓜子仁、麻仁等清洗干净备用。然后将木框放于台案上，框内薄薄地撒上一层面粉，再垫上一层芝麻。将炸好的面条均匀拌上一层糖浆后倾入框内，用手刮板铺平。表面均匀地撒上果料压平，厚度约3.5cm。放置5min左右时间，用刀切成2.5cm×5cm的长方形。冷却后用玻璃纸或糯米纸包装，按成品每千克24～40块计量装箱。

五 萨其马的风味特色

萨其马呈淡黄色至金黄色，外形规整，刀口整齐，组织疏松、不松散，口感酥松绵软，软硬适中，香甜可口，具有浓郁的桂花蜂蜜香味等特色，因此赢得人们的喜爱。

萨其马富含碳水化合物，是构成机体的重要物质，能储存和提供热能，调节脂肪代谢。脂肪能维持体温和保护内脏，提供必需脂肪酸，促进脂溶性维生素的吸收，增加饱腹感。

参考文献

方嘉沁．2018．大麦若叶牛轧糖沙琪玛*制作工艺研究［J］．现代食品，（14）：153-156．

李三宝，崔春，赵谋明．2012．沙琪玛生产配方的优化研究［J］．食品工业科技，33（09）：313-315，349．

李小娟，徐娟，张蓓，等．2019．红枣沙琪玛生产工艺研究［J］．粮食与油脂，32（01）：59-62．

梁大伟，朱萍．2011．无铝沙琪玛的研制［J］．粮食与饲料工业，（04）：23-26，30．

刘宪红，潘旭琳，田伟，等．2016．绿豆沙琪玛制作工艺的研究［J］．农产品加工，（02）：37-39

万新，周琼瑛，黎惠兰．2004．台式沙琪玛制作工艺优化研究［J］．粮油食品科技，12（04）：25-26．

王小鹤，于淼，鲁明，等．2016．响应面法优化玉米沙琪玛加工工艺［J］．食品研究与开发，37（11）：74-79．

薛红梅，刘玉美，刘晓松，等．2018．无糖苦荞沙琪玛的工艺研究及血糖生成指数评价［J］．粮食加工，43（01）：65-69．

姚科，叶玉稳，胡国华．2018．沙琪玛专用膨松剂开发中复配酸剂的影响研究［J］．中国食品添加剂，（07）：169-175．

岳佳，王若兰，李成文，等．2016．响应面法优化薏苡仁芡实沙琪玛的制作工艺［J］．河南工业大学学报（自然科学版），37（03）：65-70．

---

* 参考文献中的“沙琪玛”即为萨其马——编者。

馓 子

Sanzi

Sanzi is made by frying noodles kneaded from dough with oil. In Northern China, wheat flour is the main ingredient for Sanzi; while in Southern China, rice flour. Unique in its style and convenient and easy in practice, Sanzi is one of the traditional foods with ethnic characteristics, and an indispensable food for festivals. It has been the symbol of unity, harmony, and fraternity among all ethnic groups.

Sanzi was called "Ring Cake" and "Hanju" in ancient times. As recorded in *Jin Chu Sui Han Ji* (a book for records of customs and seasonal occasions in the Land of Chu over a year), 150 days after the Winter Solstice, Cold Food Festival comes with strong wind and pouring rain. As records in *Ye Zhong Ji* (a historical record of Yecheng, the capital city in the period of Six Dynasties) by Lu Hui in Jin Dynasty (AD 266-420) shows, the tradition that fire was prohibited in Cold Food Festival started from Jie Zitui. In the Spring and Autumn Period (770-476 BC), Jie Zitui was burnt dead in the fire in Mian mountain. Emperor Wen of Jin was so grieved that he forbade any fire at Cold Food Festival in memory of his loyalty. The Cold Food Festival became popular in the late Han Dynasty (AD 25-220) . As no fire was permitted during this festival, the ring-shaped pastry was fried in advance and served as fast food in this festival. Only served during Cold Food Festival, so it is called Hanju. As early as more than 2000 years ago, there was such sentence as "There are Junv and Mier with Zhanghuang (i.e. malt sugar)" in the *Elegies of the South · Requiem* (*Requiem* is one of the poem in the *Elegies of the South*, the first romantic poetry collection in Chinese history) written by the famous patriotic poet Qu Yuan. Lin Hong, the famous poet and gourmet in the Song Dynasty (AD 960-1279) , proved that Junv with Zhanghuang is Hanju without doubt. It's recorded in the chapter of grain of *Compendium of Materia Medica* (a book that records traditional Chinese medicine), by Li Shizhen during Ming Dynasty (1368-1644) , that Hanju is also called Sanzi, and is made with glutinous rice flour and salt. The dough is kneaded into ring shape, and then fried to create the crispy texture. Later, Sanzi was called Hanju, Junv, Gaohuan, Niantou and other names in different area and times. Legend has it that when the war broke out between the State of Chu and the State of Han, Liu Bang led the army to fight against Xiang Yu. Liu's army was well received by the people of Xuzhou for their strict discipline. To make a full meal for Liu's soldiers, the people of Xuzhou wielded their quick wits and invented the convenient and affordable pasta of Sanzi.

Sanzi was dominantly made of glutinous rice flour before the Qin and Han Dynasties, and wheat flour is used as its principal ingredient after the Sui and Tang Dynasties. The replacement of rice flour to wheat was related to the gradual promotion and use of wheat flour after Han Dynasty. At the same time, oil and milk of cow and sheep were employed in the making of pasta by ethnic groups. By the Ming Dynasty, the main ingredients of Sanzi were detailed in the *Regimen of the Song Family* (a book that records diet culture in ancient Chian) by Song Xu, including wheat flour, oil, water, and a little salt, which were almost the same as the ingredients of today's Sanzi. However, leavening agent and sugar are added in the making of Sanzi nowadays to make it more crispy and sweet.

Nowadays, fried Sanzi is found everywhere in China, especially in northwest China. It is the tradition to fry Sanzi on festivals for Uygur, Kazakh, Dongxiang, Salar and Hui ethnic groups. Thus, whenever there is a festival, each household of Hui ethnic group would fry Sanzi to entertain guests or give it to relatives and friends as a present. Once the guests came, a cup of tea and a dish of Sanzi are the best hospitality etiquette. Everyone sits together and enjoys the Sanzi while chattering with each other, immersed in the harmonious and friendly atmosphere and endless joy.

Sanzi is crispy in taste and bright yellow in color with the shape displayed in layers and melts in mouth after dipped in milk. It can be consumed dry as snack food or in soup.

馓子是用油浸面团后搓条炸制而成。北方馓子以小麦粉为主要原料，南方馓子多以大米粉为主要原料。馓子不仅造型独特，而且方便易作，是富有民族特色的传统食品之一，也是欢度节日不可缺少的食品。

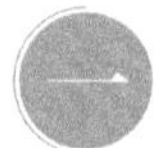

馓子的起源及发展

馓子古称“环饼”“寒具”。《荆楚岁时记》中说：“去冬节一百五日，即有疾风甚雨，为之寒食。”晋陆翙的《邺中记》有“邺俗，冬至一百五日为介子推断火，冷食三日，作乾粥，是今之糗”的记载。春秋时期，介子推焚死绵山而晋文公哀，遂立寒食不举火，以纪其忠。寒食节在汉末魏晋时非常流行，期间不能生火，于是人们便提前炸好一些环状面食，作为寒食节期间的快餐，因其为寒食节所具，就被叫作“寒具”。早在2000多年前，我国著名爱国诗人屈原的《楚辞·招魂》篇中就有“粔籹蜜饵，有餦餭兮”的句子。著名宋代词人、美食家林洪考证：“粔籹乃蜜面之干者”“蜜饵乃蜜面少润者”“餦餭乃寒具食，无可疑者”。明代李时珍的《本草纲目·谷部》记载：“以糯粉和面，入少盐，牵索纽捻成环钏之形，油煎食之。”之后因地区、时代的不同，馓子还有“粔籹”“膏环”“捻头”等别称。相传楚汉相争时，刘邦率兵与项羽作战，因刘邦的军队纪律严明而深受徐州老百姓的拥戴，为了能让刘邦的军队在行军途中吃上一顿饱饭，徐州的老百姓急中生智，发明了这种既快捷又方便实惠的面食。

秦汉之前的粔籹或细环饼是以粘稻米粉为主料，而隋唐以后的馓子大多以面粉为主料。米粉逐渐变成了面粉，这跟汉以后面粉的逐渐推广使用有关。同时出现了少数民族面食中经常采用的牛羊油、牛羊奶调面。到了明朝，宋诩《竹屿山房杂部·养生部二》的“面食制”中馓子的原料非常具体，包括面粉、食用油、水、少许的食盐。这些原料与当今馓子的用料几乎一致。只是发展至今，馓子原料调制过程中还需添加膨松剂、糖，使得馓子更加酥脆香甜。

如今，油炸馓子各地都有，西北最为常见，中国的维吾尔族、哈萨克族、东乡族、撒拉族、回族等都有节日炸馓子的习俗。每逢开斋节或其他喜庆节日，回族同胞几乎家家都要炸馓子，用来招待宾客或馈赠亲友。客人来到，一杯盖碗茶，一碟馓子端上，就是最好的待客礼节。大家坐在一起，一边“磕”馓子，一边拉家常，和睦友好的气氛洋溢其间，其乐无穷。

 馓子的逸闻趣事

馓子是宋代文学家苏东坡非常喜欢吃的食品，他曾写《寒具诗》赞美馓子的做法。

纤手搓成玉数寻，碧油煎出嫩黄深。

夜来春睡无轻重，压扁佳人缠臂金。

“点心香，月饼美，回族的馓子甜又脆。”馓子素来以股条细匀，香酥甜脆，金黄亮润，轻巧美观，而博得中外人士的赞誉。

 馓子的种类

根据地区特有的文化，馓子形成了不同的风格和特点。

滕州馓子

滕州馓子起源于19世纪末，当地主要以韩、赵、胡、薛四家馓子为主。滕州馓子历史悠久，采用的炸馓子技术是流传了两百多年的老技艺，口味正宗，外脆里香，大多为粗条馓子，少数为细条馓子，又称焦条馓子，是根据人们口味的不同，方便食用等特点为外地人而创新的一款产品。

衡水馓子

衡水油炸馓子具有香脆、咸淡适中、馓条纤细、入口即碎的特点，因售出时多扎成蝴蝶形，故又名蝴蝶馓子。衡水的蝴蝶馓子外形美观，口感颇佳。

济宁馓子

济宁王家馓子为济宁名吃，创办于20世纪70年代，创始人王宪章老先生（国家二级厨师）根据馓子的传统工艺，通过多年的探究，研制出独具特色的细条馓子，其香酥可口，色味俱佳。现王家馓子在济宁已经成为家喻户晓的地方名吃。

淮安茶馓

茶馓是用红糖、蜂蜜、花椒、红葱皮等原料熬成的水和适量的鸡蛋、清油和面，然后反复揉压，搓成粗条，捻成面团，搓成或抻成由粗细匀称、盘连有序的圆条构成环状物，放入油锅炸至棕黄色即成，是江苏省知名传统点心。其色泽嫩黄，造型秀丽，松酥香脆、独具风味。

济南馓子

济南馓子很细很长，俗称“细馓子”，焦黄酥脆，格外馋人，独特风味。通常与马蹄烧饼配套夹着吃，或放在甜沫或粥内泡着吃。

阆中馓子

阆中馓子是用盐水和面，面和好后在上面抹一层薄薄的菜油，用湿毛巾搭上，让其自然发酵，夏天发10min，冬天发60min。面发酵好后，将面盘成粗藕节般的大条，再发酵10min，再将其盘为小指头粗的小条，干米粉扑面，右手握起小条，左手捏小

条一端做纺车状运行，将小条绕成更细的面圈，然后将面圈套在两根长筷子样的竹签上，再将面圈拉细、绷直下锅用油炸，炸至面圈变硬之前，翻转竹签将其扭成梳子状，炸成金黄色时，即可起锅备用。食用时，先将油茶舀入碗中，再加上捏碎的馓子，同时加上切成小块的大头菜、捣碎的花生米、椒盐、葱花、红油等，即可上席供顾客享用。具有脆柔相融、诱人食欲、香辣爽口、暖身开胃的特点。

西宁馓子

西宁馓子其形圈圈相连，外观纤细黄亮，入口浓香酥脆，为面食中的佳品，在宴会上也扮演着重要角色。

宁夏馓子

回族馓子以宁夏回族的馓子为佳，素来以造型优美、香脆可口、金黄亮润而博得中外人士的赞誉。20 世纪 80 年代西北五省烹饪比赛表演中，宁夏馓子誉满西安。

四　馓子的制作工艺

馓子制作工艺流程

调粉 → 压面 → 饧面 → 搓条 → 盘花 → 油炸 → 成品 → 包装

参考配方

高筋面粉 250g，水 250ml，鸡蛋 50g，食用盐 6g，黑芝麻 5g，花椒 5g，茴香 2g，甘草 1g，红糖 1g，食盐 2g，发酵粉少许。

食用油（用于油炸，一般是花生油）。

操作要点

1．和面水的熬制：取水 250ml，加入花椒、茴香、甘草、红糖、食盐，煮沸，熬制 10min，冷却后除去花椒、茴香、甘草备用。

2．调粉：将 250g 高筋面粉放在和面盆里，烧一勺滚开的食用油倒入面粉中，将面烫熟后，加入 200ml 和面水一并和入。和面时，先搅拌成均匀的面絮，用手把所有的面絮揉合在一起，反复揉成表面光滑的面团。

3．压面：将面团反复揉压，中间加少许发酵粉，不断翻动，然后反复压揉，一直压到面团变得细腻、柔软、光滑、有韧性，表面起小泡，直到面团既不粘手，也不粘案板，才算是合乎标准。

4．饧面：先将面团切成面剂子（小面团），将每个面剂子反复搓揉后压成面饼，表面抹少许香油，放入大盆中用布盖压饧 10min，再拿出来搓成拇指粗的面条，叫“馓坯”，再放入盆中饧 15min 左右。这样反复揉、反复饧制成的馓子表面就不会开裂，馓子也不易断根散条。

5. 搓条：搓条是一道关键工序。将饧好的馓坯用双手搓成直径在 0.5cm 以下长约 2m 的馓条盘绕 7～8 圈，将两头对接，馓圈呈椭圆形。

6. 盘花：将搓好的馓圈套在长筷子上拉成各种形状。

7. 油炸：调整油温 150℃，将馓子放入，将筷子抽出，直到将面条表面炸成金黄色即可夹出来，放在吸油纸上吸取表面多余油分即可。

五 馓子的风味特色

馓子口感香脆，色泽黄亮，层叠陈列，轻巧美观，泡过牛奶后入口即化。馓子可以干吃，作为休闲食品，也可以泡汤食用。

馓子用油浸面搓条炸制而成，主要营养成分是脂肪及碳水化合物。另外，含有蛋白质、钙、铁、磷、钾、镁等矿物质。民间常用馓子泡汤，加以延胡索，苦楝子，对小儿小便不通有辅助作用；地榆、羊血炙热后馓子泡汤送服，对红痢不止有帮助；月子里用红糖水泡馓子，利于产后妇女散腹中之淤。

参考文献

崔庙生. 2015. 馃子，馓子，炸油香 [J]. 美食与美酒，(01)：88-95.
江涌. 2000. 馓子美名天下传 [J]. 中国穆斯林，(04)：37-38.
邝焕焕. 2015. 食材中主要化学成分对花生油煎炸过程中品质劣化的影响 [D]. 郑州：河南工业大学.
李子初. 1981. 馓子麻花与寒食节 [J]. 食品科技，(03)：26.
刘绍义. 2014. 脆如凌雪馓子香 [J]. 乡镇论坛，(36)：36.
仇杏梅. 2017. 粔籹、馓子、麻花源流考述 [J]. 美食研究，(01)：16-19.

山西刀削面

Shanxi Daoxiaomian

Daoxiaomian (sliced noodle) of Shanxi province in Northern China refers to food made from wheat flour as the main ingredient by cutting the dough into willow leaves or semi-circular and rhombus-shaped strips and then adding the gravy after being cooked. It is a kind of Shanxi traditional and distinctive local noodles, and it is also a signature of Shanxi noodle culture. It has a unique flavor and is well-known at home and abroad. Its noodles are completely sliced with knife. It is also called the "Flying Sliced Noodles" due to its excellent slicing skills. It can be praised to be the "best of world". Together with hand-pulled noodles, Boyu noodles, and Daobo noodles, they are known as "Four Major Kinds of Noodles in Shanxi".

According to the different tools, hand-made Shanxi sliced noodles can be divided into Shanxi sliced noodles made by machete and Shanxi sliced noodles made by hook knife; in light of the different working principle of the machine, the noodles can be divided into pressing-style Shanxi sliced noodles and cutting-style Shanxi sliced noodles.

Shanxi noodles have a long history and profound history. The Daoxiaomian originated from Taiyuan of Shanxi in the Yuan Dynasty (1271-1368). In the Ming Dynasty (1368-1644) established by Zhu Yuanzhang, Daoxiaomian was known as "Kan Noodles". This "Kan Noodles" was popular among small peddlers, and has been transformed into "Daoxiaomian" after many improvements. During the Kangxi's rein (1662-1722) in Qing Dynasty, Shanxi Daoxiaomian was involved with another branch–Jinxiaoer Daoxiaomian. Yu Chenglong (from

Yongning county in Shanxi province) loved to eat his hometown's Daoxiaomian. When he was an official in Zhili, he happened to walk into a newly opened Daoxiaomian shop which offered smooth and delicious noodles with lingering aftertaste. Its owner was warm and hospitable, and provided considerate service. After asking, he found out it was opened by the waiter in the noodles shop in his hometown. He was so happy that he wrote "Jinxiaoer Daoxiaomian" on the tablets. Since then the shop has enjoyed relentless guests and flourishing fame in the capital.

The cooking method of Shanxi Daoxiaomian has been recorded in *An Introduction to the Vegetarian Food* of the late Qing Dynasty. The traditional way is to hold the noodles in one hand and to slice them into the pot of boiling water with the knife in other hand. The key is that "the blade does not depart from the noodles and vice versa. Keep the arm straight hard and the hand steady. Make sure the hand and eye in a line. Without any stop, flat noodles are sliced with a normal knife; while triangular ones with curved knife." Skilled cook put dough on the top of head, with each hand holding a knife, respectively and wave their hands above their head. Modern hand-made "Flying Daoxiaomian" has "three wonders". Fast: more than 200 noodles are dazzlingly sliced per minute. Accurate: each strand of noodle is sliced into a jade plate 1.5m away. Marvelous: the cook slices the dough with his pair of hands above his head, which is the breathtaking spectacle. To enjoy the slicing performance before eating is like appreciating an artistic performance.

The legend has it that when Mongolians occupied the Central Plains, the rulers confiscated the metal utensils of every household and laid down relevant regulations that ten households used one kitchen knife. The knife is used for cooking in turn, and then returned to the custody. One day at noon, an old man's wife has made the dough and asked the old man to get the kitchen knife. But the knife was taken away. However, the old man found a thin sheet of iron by accident. He picked up and carried it in his arms. When he got home, the water was boiling. He wanted to use it in the place of knife. His wife doubted that the iron was too soft to cut the dough. The old man said, "if you cannot slice it, just chop it." A word "chop" reminded his wife. She put the dough on a wooden board and kneaded it well. With dough in her left hand, and the piece of iron in her right hand, she stood by the boiling pot and "chopped" pieces of noodles into the pot. The noodles were cooked well into a bowl and poured with brewed dishes for the old man to eat. The elder praised while eating: "Very good, so delicious. Never again you will queue up for a knife." In this way, the method of "chopping noodles" spread to tens, to hundreds, and all over the land of the Shanxi province.

With the advance of the times and the invention of a fully automatic noodle slicing robot, the making of Daoxiaomian has entered the stage of mechanization and industrialization, which plays an important role in the development of Daoxiaomian.

山西刀削面是指以小麦粉为主要食材制作面团，刀削成柳叶形或横切面为半圆、菱形的条状，经煮熟后加入面卤而制成的食品。刀削面是一种山西的传统特色面食，也是山西面食文化的招牌，风味独特，驰名中外。刀削面全凭刀削，以刀功和削技的绝妙又被称为“飞刀削面”，堪称“天下一绝”。刀削面与抻面、拨鱼、刀拨面并称为“山西四大面食”。

手工制作根据刀具的不同可分为弯刀山西刀削面和勾刀山西刀削面；机制加工按照机器制作原理不同可分为压制山西刀削面和削制山西刀削面。

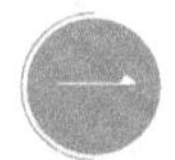

一 山西刀削面的起源及发展

山西面食历史悠久，源远流长，刀削面起源于元代山西太原。在明朝时期，刀削面被称为“砍面”。这种“砍面”流传于社会小摊贩，后经过多次改革，演变为“刀削面”。清康熙年间，山西刀削面出现一个分支——“晋小二刀削面”。于成龙（祖籍山西永宁州）自幼酷爱吃家乡的刀削面，他在直隶为官时，遇一刀削面馆开业，削面滑爽适口，回味悠长，店家热情好客，服务细微，经打问竟是老家开面馆的小二，欣喜之余泼墨题匾。“晋小二刀削面”从此食客盈门，名噪京城。

山西刀削面的做法在清末《素食说略》已有载。传统的操作方法是一手托面，一手拿刀，直接削到开水锅里。制作刀削面的技术要诀是：“刀不离面，面不离刀，胳膊直硬手端平，手眼一条线，一棱赶一棱，平刀是扁条，弯刀是三棱。”技艺娴熟的师傅，将和好的面团顶在头部，两只手中分别拿一把削面刀，在头顶上“嗖嗖”挥动。现代手工“飞刀削面”有“三绝”：快，每分钟能削出 200 根以上，令人眼花缭乱；准，1.5m 外放个玉盘，削面根根入内；奇，削面者头顶面团双手舞削，惊险壮观。吃面前，能够参观厨师削面，无异于欣赏一次艺术表演。

二 山西刀削面的逸闻趣事

（一）山西刀削面的故事

相传古时，蒙古人进入中原，将每户的金属物没收，规定 10 户共用一把厨刀，轮流使用后再交回。传说有一天，一位老婆婆和面做面条，让老汉去取刀，结果刀被别人取走。偶然之间，老汉捡到了一块铁皮，并揣在怀里。回家后，锅开得直响，忽然想起怀里的铁皮，便用铁皮代替了刀。老婆婆质疑如此软的铁皮如何能切面条。老汉说：“切不动就砍。”“砍”字提醒了老婆婆，她把面团放在一块木板上，左手端起，右手持铁片，站在开水锅边“砍”面，一片片面叶落入锅内，煮熟后捞到碗里，浇上卤汁让老汉先吃，老汉边吃边说：“好得很，好得很，以后不用再去取厨刀切面了。”这

样一传十，十传百，传遍了晋中大地。

随着时代的变迁，全自动削面机器人的问世，使刀削面进入机械化、产业化阶段，对刀削面的发展起到重要的作用。

（二）山西刀削面的赞美诗

山西刀削面顺口溜：一叶落锅一叶飘，一叶离面又出刀，银鱼落水翻白浪，柳叶乘风下树梢。

晋小二刀削面的赞美绝句：三揉四醒始为善，二尺托起白头山；五指拂出飞龙吟，一碗心香万家欢。（源自清康熙年间）

三 山西刀削面的制作工艺

刀削面制作工艺流程

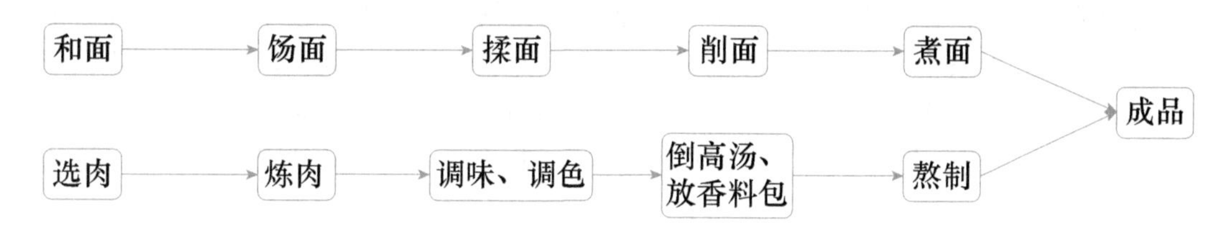

专用设备和工具

手工制作包括：勾刀板、菜刀、弯刀、勾刀等。

机制加工包括：和面机、压面机、切面机、削面机等。

原辅料

刀削面原辅料主要有小麦粉、鸡蛋、食盐、猪肉、干黄酱、色拉油、酱油、老抽、陈醋、胡椒粉、甜面酱、料酒、葱、姜、蒜、白糖、八角、黑木耳、腐竹、海带、口蘑、香菇、豆腐、西红柿、黄花菜、马铃薯淀粉、韭菜、草果、白蔻、花椒、香叶、桂皮、肉蔻、蚝油、排骨酱、辣椒及辣椒面等。

操作要点

1. 手工刀削面的制作。

1）和面：刀削面对和面的技术要求较严，水、面的比例要求准确，一般是500g面200g水，打成面穗，再揉成面团。应分3次加水，在500g面粉中应先加入水总量200g的1/2即100g；待面粉和水抄拌成穗子状或雪片状时，再加入水总量的1/4即50g；继续抄拌成大块状后

加入剩余的水；抄拌时应由外向内、由下向上，用力均匀，手不沾水，以粉推水，促进水和面粉紧密结合。

2）饧面：和好的面团应用湿布盖上或放置在密封的容器中，在15～25℃的环境下饧30min。

3）揉面：饧面后再揉，直到揉匀、揉软、揉光。揉面时身体应与案板保持适当的距离，双脚与肩同宽或呈丁字步；揉面应采用双手，右手使劲，左手辅助，用力均匀，顺着一个方向揉，以免破坏面筋网络；面团应反复揉至无干块、柔润，表面光滑，成圆柱形。揉面结束时做到面光、手净。

4）削面：削制分为弯刀山西刀削面和勾刀山西刀削面两大类型。削面时，左手托面团，应伸直小臂、手肘自然弯曲、手背与地面平行，右手持刀，从右至左、由里向外均匀地削通；削制时刀刃应紧贴面团表面，用力均匀、连贯。弯刀山西刀削面成形和规格为柳叶形条状，长20～25cm、宽0.9～1cm、厚0.2～0.4cm。勾刀山西刀削面成形和规格为菱形条状，长35～40cm，宽0.6～0.7cm，厚0.3～0.4cm。

5）煮面：至少煮沸三次，面条才能煮熟。根据个人对面条软硬程度的要求可适当加长煮面时间。

6）装碗：刀削面捞入碗中，浇上面卤，保持一定的温度。

2．卤的制作。刀削面的卤又称“臊子”、“浇头”，也称“调和”。山西刀削面的“浇头”十分考究，品种繁多，有西红柿鸡蛋酱、肉炸酱、肉丝什锦卤汤、羊肉汤、茄子肉丁卤、金针木耳鸡蛋卤等。

1）选肉：选用一品猪肉、猪排骨或新鲜鸡肉，做酱香牛肉面卤必须要选上好的牛腱子肉。

2）炼肉：将选好的肉、排骨切成小块，做大同刀削面的肉要剁成肉丁，锅中倒油，油热后放入肉块或肉馅煸炒，煸炒至肉馅变色变干即可。做酱香牛肉面卤，牛腱子肉加米酒、蚝油等调味品炒制。

3）调味、调色：在炼好的肉中依次放入葱姜蒜末、花椒面、大料面，然后喷醋，撒盐，放上好酱油等调味品，然后加入开水。亦加入各种中草药，按一定比例的调料配方去腥增香。调好后，其味美汤鲜，营养滋补，口唇留香。

4）加高汤、放香料包：骨汤是最入味的汤料，不仅鲜美，而且不膻不腥，味厚色醇。将调味好的肉加入经过长时间熬制的大骨汤，并放花椒、大料、小茴香、香叶、草果、良姜、桂皮等香料。

5）熬制：中火炖煮至肉块熟，根据口感适宜控制熬制时间。

6）成品：将煮熟的刀削面捞入碗中，浇上“卤”即可食用。

山西刀削面卤子味道咸鲜香浓，或酱香浓郁。几种常见的山西刀削面卤子配方及具体做法如下。

炸酱卤：猪前腿肉500g、葱20g、姜10g、蒜40g、色拉油100g、花椒油3g、干黄酱200g、甜面酱100g、食盐6g、胡椒粉2g、料酒40g、陈醋6g、老抽6g、花椒4g、八角4g、高汤400g。肉碎炒香后放入料酒和陈醋去腥、增香；干黄酱、甜面酱炒前用水将其稀释，入锅温度要低，将其炒散；面卤用小火熬制出酱香味。

西红柿鸡蛋卤：西红柿1000g、鸡蛋100g、葱5g、姜3g、蒜15g、色拉油50g、香油50g、食盐10g、白糖5g、老抽3g、花椒5g、八角3g、高汤1000g。西红柿粒炒香后加入高汤用小火熬制20min至味浓，香油最后加入。

小炒肉卤：五花肉500g、葱10g、姜5g、蒜15g、色拉油50g、花椒4g、八角4g、醋6g、食盐20g、料酒10g、胡椒粉1g、老抽10g、高汤1000g。猪肉在炒熟后放入料酒和陈醋去腥、增香；用小火炖制40～50min至肉香味溢出。

什锦打卤：木耳25g、腐竹25g、海带25g、口蘑25g、香菇25g、炸豆腐20g、黄花菜25g、鸡蛋20g、葱10g、姜5g、韭菜50g、色拉油50g、食盐30g、胡椒粉5g、老抽10g、花椒5g、八角3g、淀粉100g、水200g、高汤4500g。干货食材分别放入水中浸泡，待涨发后改刀成形焯水；高汤烧沸后放入食材煮制；勾芡后的卤汁呈黏稠状；韭菜、葱丝、姜丝、蒜丝在最后铺在卤汁的表面，淋入加热的色拉油。

四 山西刀削面的风味特色

山西刀削面色泽自然，长短宽窄一致，中厚边薄，棱锋分明，形似柳叶。入口外滑内筋，软而不黏，不仅可浇卤，还能热炒甚至凉拌，有独特风味。

山西刀削面以养生健身而著称。刀削面富含蛋白质、碳水化合物、维生素和钙、铁、磷、钾、镁等矿物质，养心益肾、健脾厚肠。卤中的瘦肉含有丰富的优质蛋白质和必需脂肪酸，富含提供血红素（有机铁）和促进铁吸收的半胱氨酸，对改善缺铁性贫血有好处；肥膘肉能提供高热量，并且含B族维生素、维生素E、维生素A、钙、铁、磷、硒等营养元素；大骨汤有一定的滋阴潜阳，补阴虚，清血热，养血安神的功效。

参考文献

陈雅洁. 2019. 人类学视野下的饮食文化——以大同刀削面为例［J］. 齐齐哈尔师范高等专科学校学报，(01)：63-64.

李梅，田世龙，胡新元，等. 2018. 马铃薯刀削面制作工艺优化及其品质特性分析［J］. 食品科技，43(12)：167-173.

倪子良. 2016. 山西刀削面标准化的必要性和意义［J］. 大众标准化，(07)：27-29.

孙跃进. 2016. 让山西面食走向世界［J］. 大众标准化，(07)：20-22.

王建军. 2016.《山西刀削面制作规范》解读 [J]. 大众标准化,(07):23-26.
薛海霞. 2013. 大同刀削面品牌驱动力分析 [J]. 山西大同大学学报(社会科学版),27(05):91-94.
张剑,张杰,熊增星,等. 2016. 小麦粉特性对刀削面品质的影响 [J]. 中国粮油学报,31(03):12-17,24.
郑宇. 2016. 漫谈山西刀削面标准 [J]. 大众标准化,(07):30-31.

山西花馍

Shanxi Huamo

Huamo refers to the shaped artwork made of flour after kneading and coloring. It is a Chinese folk dough sculpture. There are diversified Huamo arts in Shandong, Shanxi, Shaanxi, Henan, Gansu, Ningxia, etc., in the basin of Yellow River. In the process of spreading the art, there are also names including "Miansu" "Mianren" "Mianyang" "Huagao" "Mianhua" and "Limo". It is widely used among the folk as a token or symbol of gift for sacrifice, celebration and decoration in folk rites and festivals. As a kind of folk art, it is also inextricably linked with local customs and social lives and is highly representative in the local culture from a long-term accumulation. It embodies people's yearning for beautiful things and epitomizes the folk culture.

There are many versions of the origin of the Huamo in Shanxi. One legend can be traced back to 2000 years ago. It is said that in the Spring and Autumn Period, the prince of the state of Jin, Chong Er, fled from the country and lived a hard life. His follower, Jie Zitui did not hesitate to slice the meat on his thigh to feed him. Later, when Chong Er returned to Jin, and became the Emperor (Emperor Wen of Jin) , he rewarded the meritorious official. But Jie Zitui refused the reward and hid in Mianshan. Emperor Wen of Jin had no choice but set fire to the mountain to force him down the mountain. However, to his surprise, Jie Zitui would rather have his mother and him burnt to death than come out. In memory of Jie Zitui, Emperor Wen of Jin ordered to change Mianshan into "Jieshan" and to build a monument and a temple for him. At the same time, he also ordered that on the day of Jie Zitui's death (one day before the Qingming Festival) , no fire should be allowed on the Cold Food Festival and using fire to cook is also prohibited. Only dry food and cold food should be eaten throughout the country. In honor of Jie Zitui, local people made dough into a tomb-shaped steamed bread, and then kneaded many vivid and creative forms of birds and animals climbed on top. They are steamed out into a variety of Mianhua, and named as "Zituimo", to show their guard against the tomb of Jie Zitui.

Shanxi wheaten food have been summed up in this way: "Three best Shanxi wheaten food are as follows: mo in southern Shanxi, noodles in middle Shanxi, and pastry in northern Shanxi." Therefore, when it comes to Shanxi Huamo, people first think of the "the hometown of Huamo" – "Wenxi Huamo" in Yuncheng city, Shanxi province. Wenxi Huamo prevailed in the Ming and Qing Dynasties, and has formed a unique artistic style and a complete creative system, mainly including four series of "Huagao" "Huamo" "Mascot" and "Panding" with more than 200 varieties. The contents of the works are all-encompassing. The portraits of people, birds and beasts, or flowers and fruits are all lifelike. Because of its romantic shape, intense colors, simple and natural feature, strong local characteristics, rich and thick cultural connotations, it is renowned as a unique folk art.

In 2006, Wenxi Huamo was listed in Provincial Intangible Cultural Heritage in Shanxi. In 2008, Wenxi Huamo was rated as a National Intangible Cultural Heritage with its unique functions for eating, appreciating and etiquette. In June 2010, Wenxi Huamo stood out in Shanghai World Expo. In 2012, Wenxi county listed building the characteristic industry of "Wenxi Huamo" as a key project to promote agricultural modernization. In February 2012, at the "Wenxi Huamo Cultural Festival in China", four pieces of Wenxi Huamo works broke the records in the world's tallest, longest, largest Huamo and the largest group dough modeling, causing a great sensation. In June of the same year, they were brought to Beijing to participate in "the First International Food Competition in Beijing Olympic Park and the 6th International Competition for Oriental Cuisine Artists". Chinese and foreign gourmet commended: "Wenxi Huamo is the essence of Chinese culture and the top creation of Chinese food culture."

The shape and style of Shanxi Huamo vary a lot. Some are famous for being bold and rough; some, elegant and noble; others, beautiful and delicate. Their common characteristics are: beautiful in shape, concise in composition, diverse in form, and rich in content, all deeply favored by people. It is especially famous for "three high quality and four uniqueness". "Three high quality" refers to high-quality wheat (planted in quality soil with a big temperature difference between day and night and long frost-free period), high-quality flour (with white color, strong gluten, high flour processing degree and no additives), high-quality water (deep underground water 300 meters below the surface, sweet and clear) . "Four uniqueness" refers to unique making method (self-made yeast, secondary fermentation, hand kneading, accurate control of the fire and eight procedures), unique nutrition (higher content of sugar, starch, protein, etc. than similar food), unique shape (the shape is like a mountain peak, dragon and phoenix, tiger, flower, peach, etc.) and unique taste (chewy, fragrant, slightly sweet, pure, and palatable).

花馍指面粉经过捏塑上色之后制成的造型艺术品，是中国民间面塑品，在我国黄河流域的山东、山西、陕西、河南、甘肃、宁夏等地都流传着各具特色的花馍艺术，流传过程中也有了“面塑”“面人”“面羊”“花糕”“面花”“礼馍”等叫法。民间花馍的应用极其广泛，参与各种民间礼仪活动中，如祭祀、祈祷、喜庆、馈赠、装饰等。花馍本身是一种民间艺术品，又与各地风俗人情有着千丝万缕的联系，是一种长久积淀而成极具代表性的地方饮食文化，寄托了人们对美好事物的向往之情，同时也是信仰和民俗文化的缩影。

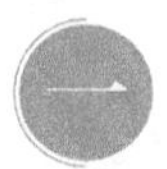

一 山西花馍的起源及发展

关于山西花馍的起源，有多种说法。相传春秋时代，晋国公子重耳逃亡在外，生活艰苦，跟随他的介子推不惜“割股奉君”。后来，重耳回到晋国，做了国君（即晋文公），封赏有功之臣。唯独介子推“不言禄”隐于绵山，晋文公无计可施，只好放火烧山，逼其下山。谁知介子推母子宁愿被烧死也不肯出来。为纪念介子推，晋文公下令将绵山改名为介山，并修庙立碑。同时，还下令在介子推遇难的这一天（清明节前一天）“寒食禁火”，举国上下不许烧火煮食，只能吃干粮和冷食。当地的百姓为了纪念介子推，用面粉捏成坟头形状的大馒头，然后再捏一些形态各异栩栩如生极富创意的飞禽走兽爬在上面，蒸出后成为花样繁多的面花，并且起名“子推馍”，表示对介子推坟墓的守护。

坊间对山西面食曾有高度赞赏：“三晋面食好，晋南的馍，晋中的面，晋北的糕。”因此一提到山西花馍，人们首先想到的是“花馍之乡”——山西省运城市闻喜县的“闻喜花馍”。闻喜花馍盛行于明清，已形成独特的艺术风格和完整的创作体系，主要包括“花糕”“花馍”“吉祥物”“盘顶”四大系列200多个品种。作品内容包罗万象，不论人物肖像，还是飞禽走兽，或是花卉果实都栩栩如生，因其造型浪漫，色彩炽烈，淳朴自然，地方特色浓郁，文化内涵丰富而厚重，可谓民间艺术奇葩一朵。

2006年，闻喜花馍入选山西省级非物质文化遗产名录；2008年，“闻喜花馍”以其独有的食用、观赏、礼仪三大功能，入选国家级非物质文化遗产名录；2010年6月，闻喜花馍在上海世博会上大放异彩。2012年，闻喜县把打造花馍特色产业列为推进农业现代化的重点工程。在“中国闻喜花馍文化节”上，四件花馍作品分别创世界最高、最长、最大花馍、最大面塑群等记录，轰动效应极大；同年6月，赴京参展“首届北京奥林匹克公园国际美食大赛暨第六届东方美食烹饪艺术家国际大赛”。中外美食家称赞：“闻喜花馍是中华文化精髓和中国食文化顶尖创作。”

在闻喜，人们常称“有馍就有事，有事就有馍”。过去每逢过年过节或者哪家娶媳妇、嫁闺女、孩子过满月等，村里的妇女们总是热热闹闹的围在长长的案板跟前，一

边拉家常，一边蒸花馍，那情景真是热闹喜庆。花馍就是人们对艺术的传承和创造，也是对祖辈情感的寄托。

 山西花馍的逸闻趣事

（一）羊羔儿馍

十里风俗不一般，山西各地的面塑形式和风格各异，其中要数霍州市较为突出。当姑娘出嫁的第一年，娘家必须在七月十五中元节给女儿送数百个“羊羔儿馍”。之所以叫“羊羔儿馍”，一说古时羊即祥，取吉祥之意。又说羊羔儿活蹦乱跳，取其活泼可爱之意。还有一说，从前一忤逆之子，苛刻老母，屡教不改，舅舅特意领他一起去放牧，借机劝教。其间，他见到羊羔跪在地下吃奶，感到惊奇，便问原因，舅舅趁势给他讲了乌鸦反哺、羊羔跪乳，不忘父母养育之恩的道理，孩子幡然醒悟，改邪归正。为使后代永记羊羔跪乳，便制做面塑“羊羔儿馍”，所以“羊羔儿馍”又有教育后代莫忘父母哺育恩情之意。

（二）山西花馍之最

最高花馍——龙王神像面塑

由闻喜县畖底镇众村民制作的龙王神像面塑，高 14m，创造了世界最高的面塑纪录。龙王神像面塑用面粉、钢筋和泡沫三部分组合而成，用泡沫 45m^3，用面粉 1000 余千克，钢架结构有 3t 之多。历时半个月时间，20 多个人才把它制作完成。

最长花馍——神龙面塑

神龙面塑曲线长 30.05m，创世界最长的面塑纪录。神龙面塑是由礼元镇阜底村制作，共用白面 100 多千克，用人力每天 80～100 人，制作时长达半个月。

最大花馍——龙腾盛世大花馍

花馍高 4.06m，直径 2.01m，近二十层的馍馍凝聚了民间艺人的独特设计。最上层是天安门城楼的造型，下面盘着九条世龙，底座的第二层是象征着 56 个民族团结兴盛的向日葵，第三层是百花争艳，第四层是四季发财，第五层是五谷丰登。

最大面塑群——裴氏宰相将军群塑

由闻喜县礼元镇裴柏村村民制作的裴氏宰相将军群塑，排成长长的两列，一眼望不到边。裴氏宰相将军群塑共有 113 尊闻喜历史人物花馍造像，平均身高为 1.8m，创世界最大的面塑造像群像纪录。原材料就地取材，用龙骨搭架，蒸熟以后用彩绘把像勾勒出来，共用白面 2t。

最大建筑造型面塑——鹳雀楼面塑

2016 年 8 月，在太原举办的“首届中国山西面食文化节”上，由闻喜县卫嫂花馍制作的鹳雀楼面塑，高 7.39m、重 3650kg，按实物以 10：1 比例制作。设计为三层四檐结构，骨架外层包裹了一层厚厚的面塑，檐边以五颜六色的面花装饰，十分喜气，

被誉为最大的建筑造型面塑。

（三）花馍的美好寓意

在花馍造型文化中常常取谐音法来表现吉祥观念，就是选取和吉祥寓意发音相似的字来取得一定的修辞效果。例如，“莲”和“连”，“鱼”和“余”的谐音，莲花、鲤鱼进行组合，就象征着“连年有余”；“羊”和“祥”，用羊的形态来表示吉祥；“鹿”和“禄”，“冠”和“官”，鹿和桂冠组合来表示“高官厚禄，加官进爵”；“枣”和“早”，“桂”和“贵”，枣和桂枝组合寓意“早生贵子”；还有“猫”和“蝶”组合寓意“耄耋”长寿；“蜂”和“猴”寓意“封侯”；荔枝表示聪明伶俐；牡丹表示富贵。这些谐音的寓意久而久之已经成为一套固定搭配，人们看到表面的图像就能想到其中的寓意。

山西花馍的类型

千姿百态、琳琅满目的花馍，不但历史悠久，久负盛名，自古以来深受人们的喜爱，而且在制作和用途上也十分讲究。如闻喜花馍的种类总体分为观赏型和实用型两大类，它与民间的重大礼仪活动密不可分，是节日、婚嫁、寿诞等大事少不了的供品。

花馍按功能可分为节日祭祀类、生日祝寿类、盖新房立柱上梁类、定亲结婚类、丧葬礼仪类；按造型可分为糕类、馍类、吉祥物类、盘顶类四个系列。其中“花糕”是面塑之最，呈磨盘状，共四层，其代表有“五谷丰登节节高”“五福捧寿”“龙凤糕”“九凤朝阳”等。“花馍”是面塑系列中的主体，应用最为广泛，在民俗活动中无一不用到花馍，是用半球形、桃形、鱼形等面团作主体，上面扦插各种动植物和人物，主要有“大烧馍”“对对馍”“枣花馍”“馄饨馍”“石榴馍”“龙凤配”。

随着社会的发展，花馍这门独特的文化艺术更加焕发出绚丽的光彩。根据不同的实际应用，花馍大致归纳为以下几种类型。

婚姻习俗花馍

男婚女嫁乃人生一大喜事。花馍从古至今是山西婚庆嫁娶礼仪中的主要文化，民间传统十分讲究。两姓联姻，花馍必不可少，订亲时男家必备“龙凤糕”和“对对大花馍”，女家也备有“大烧馍”和“馄饨馍”，寓意“天赐良缘、荣华富贵”。迎亲时有专人捧端“上头糕”，糕的中间是一个大石榴，左边是凤，右边是龙，石榴前有一缕五彩线。石榴要在送到娘家迎亲时拔下来，意味着早生贵子。周围用面做成的花鸟，底座用厚厚的枣糕做成。待新娘出门时，妈妈要将“上头糕”的根部一段裁下来，让一对新人带走，意为女儿应在婆家扎根，其余切成片分送给家族与邻居，表示喜鸟飞远了。

生育习俗花馍

孩子出生和过满月时要举行隆重的祝贺仪式，要蒸制“枣长花馍”，寄托着父母对儿女的殷切期望。亲朋好友要送“大老虎”、“老虎腿”等面塑，以示祝福，寓意“岁

岁保平安、虎虎添生气”。孩子周岁要吃“鱼花馍”，意为一岁有鱼，岁岁有余；孩童十二岁时要吃“项圈花馍”。

上梁乔迁花馍

自古以来，在上梁和乔迁时，民间就盛行庆典和祭祀的礼仪，蒸制“上梁糕”以祭祀神灵；蒸制“九狮拱菊糕”、“桃型馄饨大花馍”等，寓意“大吉大利、万事如意”。如今伴随着时代的发展，为了更好地表达人们的美好心愿和祝福，人们将民间流传的“福寿禄”三吉星，精心制作成“福寿禄”大型花糕，寓意“五福临门、长命百岁、高官厚禄”。

开业庆典花馍

开业庆典不但要择定吉日，店主还要聘请面艺技人蒸制一个特大“花糕”和八个大花馍，意为“大吉大利、万事如意”“岁岁登高、年年鸿发”“前程似锦、宏图大展”。亲朋好友前去恭贺，都会敬送一个“吉祥”面塑大花篮，寓意着“兴旺发达、财源滚滚”。有的亲友还会蒸制“双双对对大花馍”，以示“吉星高照，财运亨通”。

祝寿贺禧花馍

人逾花甲之年，儿孙晚辈们要怀着孝心为其祝寿贺禧，以此秉承和弘扬德孝家风。祝寿礼俗中的花馍表达的主题是“孝”，有“五福捧寿”“八仙庆寿”“五女拜寿”等，更多的是“九狮拱菊”，谐音“九世共居”，寓意“四代同堂”“五世其昌”，将寿诞文化表达得淋漓尽致。

祭祀拜祖花馍

在民间祭祀拜祖礼仪文化中，人们都要制作精美的花馍作为祭品，以表达心意和彰显隆重之礼。“枣山”花馍在民间祭祀神灵中，寓意“早生贵子、大富大贵”；清明节扫坟祭祀中用的“飞燕”花馍，表示春燕飞来，阳光明媚，万象更新；农历十月初一，家家蒸制大花馍进行祭祀，意为去世的亲人送饭添衣度寒冬。

岁时节庆花馍

岁时节庆是我国传统民俗的重要礼仪。过年要吃“枣花馍”“大枣山”“枣篮馍”“元宝馍”，象征着在新年里能招财进宝，寿福安康；吃“镇宅吉祥老虎”、“狮子”等花馍，以寄托平安吉祥。元宵节要蒸“馄饨花馍”，喻为一家老小皆平安，四邻和睦乡情浓。清明时节的“桃花馍”，意在传递春的信息。五月端午节家家蒸制“虎头花馍”，寓意驱邪避毒、消灾免祸。六月六做“莲花馍”，象征着品行高洁、清白一生。中秋节的“糖枣月饼馍”，以示祭月拜祖，共庆团圆。九九重阳节的“节节糕”“寿糕”“菊花馍”，不仅意含“五谷丰登节节高”之意，更表达对长辈们的孝心和祝福。

面塑工艺花馍

作为花馍文化艺术中的精髓，面塑花馍的品种繁多，琳琅满目，既有食用性，又具观赏性，制作非常讲究，且题材十分广泛。有些花馍还加进了传说故事、神话以及戏曲人物等，增加了知识性和趣味性，诸如浮雕式的“嫦娥奔月”“五龙碑雕”“八仙过海”“麻姑献寿”等系列花馍艺术，表现后稷出生的“龙生虎养雕打伞”，表现河津龙门神话的“鲤鱼跳龙门”，戏曲故事“苏三起解”“卖水”“舍饭”等。令人叹为观止，其工艺和收藏价值十分可观。

花馍筵宴

闻喜县在立足于挖掘、保护闻喜花馍这一“国家级非物质文化遗产”的基础上，不断开拓市场，潜心研发出富有营养，又有观赏性的“花馍筵宴”。主要推出“龙王宴”“关公宴”“宰相宴”“状元宴”“功臣宴”“寿诞宴”“满月宴”“鸳鸯宴”等十多种花馍宴。一桌桌构思绝妙、制作精美的馍宴，一个个栩栩如生、造型逼真的面塑，展现了博大精深的花馍文化艺术气氛和“一帆风顺、四季平安、万事如意、招财进宝、步步登高”的美好意境。

四 山西花馍的制作工艺

（一）原料介绍（以闻喜花馍为例）

闻喜花馍的制作食材主要有北垣麦子和北垣水。

闻喜花馍之主料——北垣麦子

闻喜花馍选用闻喜北垣产的最优质小麦磨成的面粉。北垣地处峨嵋岭腹地，此地盛产的小麦生长周期长，粒大饱满，麦质优良。用此地小麦面做出的馍，清香筋道、营养丰富，吃在嘴中，甜中带香。

闻喜花馍之主料——北垣水

制作闻喜花馍不仅要用北垣麦子，同时要用这里的水和面，因为北垣地势较高，水位深，300m以下的深层水，甘甜清澈，水质好。

（二）山西花馍的制作（以闻喜花馍为例）

山西花馍制作工艺流程

凝水→箩面→制酵→揉面→捏形→醒馍→蒸制→着色→插面花

操作要点

1．凝水：用筛子、簸箕筛小麦，簸净杂物，再用水淘洗，而后装进竹篮或布袋放在阴凉处凝水，存放一日，便于麦粒皮与麦粒分离。

2．箩面：专门选取精粉，再用最细的丝制面箩过一遍，去掉其中的粗粉粒和碎麦皮。

3．制酵：闻喜花馍用酵面发酵，而不用发酵粉或碱面发酵。酵面是用玉米面做原料，选取少量酵母掺进一定数量的笼面，用水搅和后放在温度适宜的地方进行初期发

酵；需提前一夜发“酵水”。“笼面”指自家制作的老酵母，就是在前一次做馒头的时候会留下来一块面，装在碗里放在干净的地方，等到下次再用时把它和刚和好的面揉到一起，面团就会自动醒发，等醒发好了后，再留下来一块，以后再用。

4．揉面：酵面制好后，按蒸馍量的多少取用，掺进等量面粉，用温水和匀后，进行揉制，至少要反复揉八遍，以瓷光有筋、软硬适度为标准。软硬度很重要，硬了蒸不透，易裂口，软了易变形。可在和面时在面粉中掺入适量的醋，可使蒸出的花馍白亮、光滑。有的加糖水或牛奶、蜂蜜，可使花馍不易干裂。

5．捏形：可利用常用的擀杖、筷子、竹签、剪刀、梳子、菜刀等工具，通过切、揉、捏、揪、挑、压、搓、拨、按、卷等十种制作手法，依据提前确定好的花馍品种，主体配件分开捏制或一体捏制。

6．醒馍：把捏制好的面塑品放在热笼圈里，再在笼圈中间放一碗热水，用棉褥盖严，保持适当温度和湿度，使花馍制品不变形、不干裂、湿润光滑。待馍发虚后，即可上笼。

7．蒸制：花馍上笼蒸制时需注意水开后上笼，急火上气，笼圈要封严，注意放气，大小花馍分开蒸，且放气后及时出笼等。

8．着色：花馍出笼后，趁热用食用色素点染、描绘，可使颜色艳丽，不易褪色。

9．插面花：待花馍晾凉后，根据形状总体的需要，用竹签插上陪衬的面花，组合成一个完整的花馍。

五 山西花馍的风味特色

山西花馍的造型风格各异，有的以粗犷豪放见胜，有的以典雅高洁赢人，有的则以优美细腻著称。其共同特点是造型美观，构图简练，形式多样，内容丰富，尤其以“三优四特”著称。“三优”即优质麦（土质性能好、昼夜温差大、无霜期长），优质面（色白、筋强、精度高、无任何添加剂），优质水（300m以下地下深层水，甘甜清澈）；“四特”即制作工艺独特（自制酵母、二次发酵、手工揉捏、把握火候、九道工序），营养独特（糖、淀粉、蛋白质等含量高于同类食品），形状独特（形如山峰、龙凤、老虎、花、寿桃等），口感独特（筋道耐嚼、浓香微甜、清纯适口）。

花馍以面粉经发酵制成，主要营养素是碳水化合物，是人们补充能量的基础食物。

中医认为，花馍具有一定的食疗作用，有助于养胃消胀。

参考文献

白晓旭. 2018. “有馍就有事，有事就有馍”：闻喜花馍的饮食人类学研究［D］. 呼和浩特：内蒙古师范大学.
白云峰. 2009. 花馍的隐喻与礼俗的现实［D］. 沈阳：辽宁大学.
郭阳. 2011. 山西面塑“花馍”的造型艺术与文化特征［J］. 包装世界,（03）: 110-111.
姜湧. 2015. “山西花馍”作为地方美术课程资源开发的研究［D］. 重庆：重庆师范大学.
孔璐. 2016. 探析民间花馍中的设计美感［J］. 大众文艺,（18）: 45.
聂巧荣. 2019. 山西“闻喜花馍”系列包装设计研究［J］. 中国包装, 39（05）: 27-29.
乔呵呵. 2018. 山西花馍的功能研究——以山西马村为例［J］. 才智,（08）: 178-179.
岳树明. 1997. 华夏一绝，山西花馍［J］. 粮食问题研究,（07）: 40-41.
张咪咪. 2018. 浅析山西闻喜花馍的发展［J］. 度假旅游,（12）: 204.
张鹏飞. 2016. 论山西闻喜花馍造型与色彩的象征性［J］. 艺术科技, 29（11）: 236-237.
赵芳. 2017. 民间花馍的发展历史［J］. 美术教育研究,（05）: 42.

凉皮

Liangpi

Liangpi, also known as Niangpizi in Chinese, evolves from cold noodles and is famous for its color, taste and aroma. It is an important delicacy in West Shaanxi and is reputed to be one of the "Four Snacks of Unique Flavor" in Hanzhong. It tastes cool, soft, sour and spicy.

Liangpi originated from 221 BC to 207 BC in the heyday of Qin Dynasty, which was recorded in the *Records of Chang' an County* and *Records of Liuba County.* Liangpi can be regarded as the pioneering food at this period.

Liangpi is a famous snack of Qin town of Hu county, Shaanxi province. On May 11, 2007, the making method of the traditional cold noodles, Qin Town Liangpi, was listed as one of the first batches of Intangible Cultural Heritage in Shaanxi province, and was exhibited at the Intangible Cultural Heritage Achievements Exhibition in Shaanxi province on June 6, 2007.

Qishan rolling dough is a very famous snack in Shaanxi, especially in the west area. Qishan rolling dough was identified as a famous Chinese snack in 2011.

Liangpi has always been popular in Northern China, and has gained popularity in the southern part of China and abroad in recent years. "Famous Snacks in Northwest China" dining car was run by Xie Yunfeng at the gate of Columbia University, and "Famous Food of Xi'an" chain stores were in business in midtown of Manhattan, which have made Liangpi the food fashion in New York City. What's more, the "Famous Food of Xi'an" has opened a number of chains in New York, and it has been listed as one of the best fast food in New York by ZAGAT (authoritative food survey organization of the US).

Once there was an emperor who was tired of eating exquisite delicacies, so he had no appetite for the food delivered by the imperial kitchen. All his wives and servants in the palace were in panic for it. A concubine, who came from today's Shaanxi, Gansu or Ningxia province, helped her mother with the Liangpi business at an early age. She was considering that Liangpi might suit the Emperor's taste. Therefore she asked the chief manager of the imperial kitchen to prepare related ingredients and she made several bowls of Liangpi all by herself. Then she sent the emperor one bowl of Liangpi as a try. When he was bored with reading the documents handed by the officials, the emperor was attracted by the spicy smell and asked the eunuch for it. The light blue translucent Liangpi noodles were set on the bottom of the bowl, with shredded cucumber, coriander and chili on the surface, which can greatly arouse one's appetite. The Emperor tried and enjoyed it at his first bite. Then he ate several bowls of Liangpi without the elegant bearing, and even asked for more. The eunuch was worried about that eating too much Liangpi would be bad for his health and suggested asking for the imperial doctor's advice. The imperial doctor was called in and he asked for the ingredients of Liangpi and then answered that Liangpi was made from root and stem of herbal plants, which might make people energetic after eating; cucumber and coriander were all beneficial food. Often eating Liangpi had no harm to health. However, don't eat too much at once. The emperor ordered that a bowl of Liangpi should be served every day and promoted the concubine to be the imperial concubine. From then on the name of Guifei Liangpi was spreading and became the delicacy popular among the imperial officials.

Liangpi noodles are white, thin, soft, and smooth and they taste spicy and oily but not greasy. Added the specially made chili sauce, their aroma will linger in your mouth after eating.

凉皮，又称“酿皮子”，因原料、制作方法、地域不同，有热米（面）皮、擀面皮、烙面皮、酿皮等。口味有麻辣、酸甜、香辣等各种口味。凉皮从冷淘面演变而来，凉爽可口，绵软润滑，酸辣开胃，是关中西府的重要名吃，被誉为汉中风味小吃“四绝”之一，以“白、薄、光、软、筋、香”而闻名。

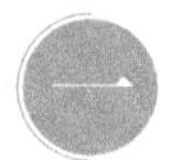

一　凉皮的起源及发展

（一）凉皮的历史渊源

凉皮是流行于我国北方的一种传统美食，起源于秦朝（公元前 221～前 207 年），距今已有两千多年历史，由陕西汉中一带推广至全国，并传到了国外。凉皮以“白、薄、光、软、筋、香”而闻名，深受国人喜爱，被誉为“粉食时代”的先锋食品。

凉皮最早被称为“面皮子”，《长安县志》和《留坝县志》中均有记载。相传，秦始皇当政时期，有一年汉中地区适逢大旱，沣河缺水，户县秦镇一带稻子干枯，碾出的大米又小又干巴，根本没法向皇帝纳贡。当时有位农民将大米碾细过筛，把米粉加水稀释调成糊状，倒入甑子蒸熟，切成细条，筋丝柔韧，软而不断，恰像皮条，名为“面皮子”，然后调入色泽红亮、辣香诱人的油泼辣子及盐、醋、酱、芝麻酱等，再佐以豆芽、芹菜，清香扑鼻，酸辣爽口，邻里乡亲品尝后念念不忘，个个称奇。有官员将此“面皮子”呈献给秦始皇，秦始皇吃了美味可口的“面皮子”，颇感稀奇，倍加赞赏，并令汉中地区每年进贡。此后，凉皮经过 2000 多年的不断发展改良，逐渐形成了汉中米皮、秦镇米皮、岐山擀面皮、西安麻酱酿皮等不同地域风味的凉皮，口味亦有麻辣、酸甜、香辣等。

（二）凉皮的发展

陕西秦镇米皮入选文化遗产名录

秦镇米皮是陕西户县秦镇的著名特色小吃。2007 年 5 月 11 日，凉皮的经典之作秦镇米皮制作工艺收入陕西省第一批非物质文化遗产名录，并于 2007 年 6 月 6 日，在陕西省非物质文化遗产保护成果展上展出。

中华名小吃——宝鸡岐山擀面皮

岐山擀面皮是陕西非常有名的小吃，尤以陕西西府地区为最佳，宝鸡岐山擀面皮于 2011 年被认定为中华名小吃。

西安小吃走红纽约，老外也爱肉夹馍和凉皮

凉皮在北方都是非常受人们喜欢的，近年来凉皮的火热劲头也传到了南方各省市，并传到了国外。哥伦比亚大学门口谢云峰的“中国西北名吃”餐车，纽约曼哈顿中城的“西安名吃”西安小吃连锁，使凉皮成为了纽约城的新食尚。

二 凉皮的逸闻趣事

描写凉皮的赞美诗很多，现摘几首供大家欣赏。

秦镇米皮：

秦渡桥头十二郎，天教巧手弄蒸尝。
细裁玉纸薄筋软，一点香红动始皇。

岐山擀面皮：

小吃谁将进御京？繁难八序品民情。
光筋薄软村香味，赢得康熙钦赐名。

汉中米皮：

水软山温细腻多，汉中米食亦柔和。
且将千种缠绵味，调到舌间凭叹哦。

西安麻酱酿皮：

精酿慈心细细抡，餐餐美食谢娘亲。
三番尚恐儿犹饿，还把当年老碗寻。

三 凉皮的分类与流派

（一）凉皮的分类

根据原料和加工工艺的不同，凉皮可分为面皮、擀面皮、米面皮。

1．面皮：以小麦粉、小麦淀粉、水为原料，经搅拌、和面、熟制成形、冷却、切制（或不切制）而成。呈白色或淡黄色，色泽均匀一致；具有小麦粉香味，无霉味；经切制的面皮呈条形，均匀完整，排列整齐、略有弹性，无糊面及明显并条；未经切制的面皮表面光滑、薄厚均匀，无糊面。

2．擀面皮：将小麦粉、小麦淀粉、水按一定比例倒入洗面机中进行均匀搅拌、发酵、加热熟化、成形、冷却、刷油、切制而成。呈淡黄色、灰白色或褐色，色泽均匀一致；具有小麦粉香气，略酸，口感筋道，无霉味及其他异味；薄厚均匀，外形完整。

3．米面皮：将大米浸泡、磨粉、加水搅拌，经笼屉蒸熟后，冷却、刷油、切制而成。呈白色，色泽均匀一致；具有大米特有的滋味、气味，无异味或微酸味；呈条形，均匀完整，排列整齐，略有弹性，无糊面及明显并条。

（二）陕西凉皮的四大流派

陕西的陕南、陕北、关中都有凉皮，但因做法、吃法、调料、用料不完全相同形成四大流派。

1. 汉中米皮：汉中属陕南，盛产大米，用大米面做面皮由汉中人首先发明，历史悠久。汉中米皮口感柔嫩、切得偏宽（4cm 左右），更加偏重于调料的制作，且调料偏重于香、辣，醋用得较少；一般热吃。

2. 秦镇米皮：秦镇米皮是陕西户县秦镇的著名特色小吃。辣椒油是秦镇米皮调料中最关键的，秦镇米皮好吃与否主要取决于辣椒和辣椒油。辣椒油是将辣椒面放入上等的油中，加入花椒、茴香、大料等小火反复熬制而成。秦镇米皮比汉中米皮口感要稍硬，更适合年轻人、中年人食用。

3. 西安麻酱酿皮：陕西农村用小麦面蒸的凉皮，一般叫酿皮。其吃法和做法与米面皮无差异。但在有些人，除了放醋、盐、辣椒油外，还要放芝麻酱，吃来又别有风味，人们把这种酿皮叫作西安麻酱酿皮。

4. 岐山擀面皮：岐山擀面皮是关中西府小吃代表之一，也是西北最具民族风味的食品之一。具有筋道、柔软、凉香、酸辣可口的特点。

四　凉皮的制作工艺

（一）原辅料介绍

凉皮的主要原辅料有高筋面粉、醋、酱油、油泼辣子、食用油、豆芽、芹菜等，因人口迁徙与文化的交融，也有加入黄瓜、胡萝卜、香菜、蒜、芝麻酱等配料，赋予凉皮更丰富的味道。

凉皮之主料——高筋面粉

高筋面粉颜色较深，有较好的活性且光滑，手抓不易成团状，且蛋白质含量高，筋度强，比普通面粉做的凉皮更白净，使凉皮更筋道、爽滑，口感好，保证了凉皮“白、薄、光、软、筋”的特点。

凉皮之滋味点睛——油泼辣子

凉皮香不香关键在辣椒。凉皮的油泼辣子选用优质辣椒，并配以多种香料，采用独特的方法制成。将辣椒、花椒、茴香等香料碾细加入油中，上火加热反复熬制而成，其色泽红亮，辣香诱人。油泼辣子红艳如火，再佐以豆芽、芹菜等辅料，使凉皮黄绿相间，堪称绝配。

油泼辣子的具体做法包括：碾辣面→油泼辣子→激香→润色。

1. 碾辣面：取当年的干红辣椒，放入大铁锅中文火翻炒至辣子的辣味和香味充分散发，辣子呈鲜红发亮时，倒入碾子中碾成合适粗细的辣椒面备用。

2．油泼辣子：取辣椒面100g放入容器中，加入食盐和五香粉各10g搅匀。锅内倒入食用油200g，加热后关火静置，到不冒油烟时分成三次倒入装辣子的容器，即油泼辣椒面，每次都要搅动均匀以免油泼不均匀。

3．激香：倒完油后搅动辣子到不冒泡时，倒入岐山用玉米、小麦、高粱等酿制的粮食醋10ml，马上搅动辣子，使辣子再次沸腾冒泡腾起香气。激香后的辣子色泽鲜红油亮，散发出浓浓略微带点酸味的醇香气味。

4．润色：激香后再等辣子不冒泡的时候，加入砂糖10g（蜂蜜15g），搅拌均匀，使白糖充分利用辣子的余热溶解于油泼辣子中。润色后的油泼辣椒红润厚重，辣子油较为黏稠。

凉皮之灵魂——醋

醋是凉皮的灵魂，醋赋予了凉皮特有的酸爽的味道，食用时促开胃。凉皮中的醋的核心在于熬醋，而熬醋的作用在于去掉醋本身的涩味。醋多选择岐山头道醋或熏醋，加水烧开，转入小火，并加入生姜、花椒、大蒜等配好的调料熬制。

凉皮之辅料——豆芽、芹菜、香菜、黄瓜、蒜泥、胡萝卜等

1．豆芽增加凉皮的口感，丰富凉皮的营养，预处理，洗净，焯水，待用。

2．芹菜作为辅料，在凉皮里起增进食欲、丰富凉皮口感、增添色泽的作用。芹菜先预处理，洗净、切段，焯水，待用。

3．香菜可增加凉皮的风味，赋予凉皮清香。预处理，洗净、切碎，待用。

4．黄瓜味甜，胡萝卜质脆味美，作为凉皮辅料，增加凉皮的脆感，丰富凉皮色泽和营养。预处理，洗净，切丝后待用。

5．蒜泥是将大蒜去皮捣碎成泥状，作为凉皮辅料，增加凉皮的风味。数种蔬菜辅料的调配，使凉皮营养更加丰富，色、香、味俱佳。

（二）凉皮的制作

凉皮制作工艺流程（以擀面皮为例）

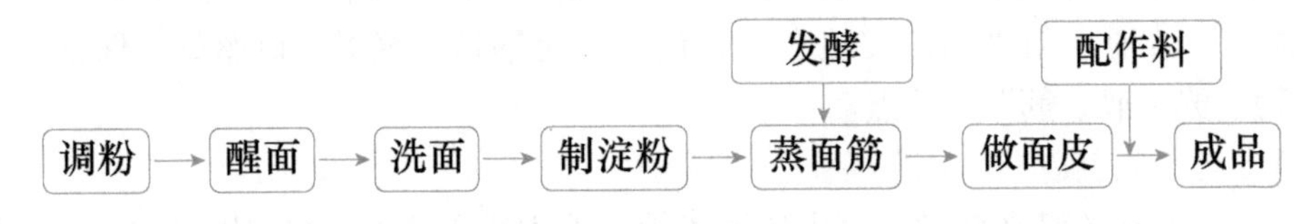

参考配方

高筋面粉3000g，绿豆淀粉900g，食盐90g，可制作面皮30张。

操作要点

1．调粉、醒面：向高筋面粉、绿豆淀粉内加入食盐，再加入适量水和成干湿适中的面团。面团先盖上保鲜膜或湿布静置30min。

2．洗面：以清水揉洗面团，将粉浆澄出，再冲洗，反复冲洗澄出，直至水变清，没有沉淀物出来，面筋干净有劲为止。

3．制淀粉：将澄出的粉浆在桶内静置4～5h。

4．制作面筋：在面筋中添加适量酵母，揉匀后盖好发酵至1倍大。锅中倒入清水煮沸，将洗好的面筋揪成条状缠成比拇指稍粗的棒状，放入锅中煮45min，用漏勺拎干水捞出，撕成较小条片状放盘中备用。或将发好的面筋上锅蒸15min，蒸好的面筋冷却后切段备用。

5．做面皮：面浆水沉淀后舀去表面的清水，剩下的粉浆用勺子搅匀；在蒸盘底刷上熟油，再舀入适量粉浆，热水上锅蒸约3min，面皮鼓起呈透明状时立刻取出，趁热取下，晾凉。

6．配料：准备拌凉皮所用的配料，胡萝卜去皮刨丝，黄瓜刨丝，香菜洗净切段，蒜剁成蓉；起油锅把花生煎炸至裂开时控油取出，撒上盐放冷却；水烧开把胡萝卜丝下锅焯20s后捞出过冷水。

7．调味：面皮一张一张分开，切成蒜苗叶形的条，加入面筋、花生、胡萝卜丝、黄瓜丝、香菜、蒜蓉；调入适量盐、生抽、蚝油、白糖、芝麻油，最后加入油泼辣子和特制的醋搅拌均匀即可。

五　凉皮的风味特色

凉皮色白薄软、晶莹透亮、筋韧不断、口感润滑、香辣味美、油而不腻、酸辣可口、爽口开胃、饶有风味。配上辣椒油食后，唇齿留香，回味无穷，是人们普遍喜爱的食品。

《本草纲目》中“米能养脾，麦能补心”。凉皮性平、味甘，有温肺、健脾、和胃的功效，是不可多得的地方小吃。

参考文献

春晨．1999．大米凉皮的制作［J］．新农村，（03）：20.

李鸣．2002．用面粉加工可口凉皮［J］．南阳农业科技，(04)：7．
暖暖尚．2015．苋菜汁粉色凉皮［J］．饮食科学，(05)：64．
权枝英．2016．挡不住的魅力——西安凉皮［J］．孔子学院，(03)：20-21．
陕西省卫生和计划生育委员会．2016．DBS61/0011-2016，食品安全地方标准 凉皮、凉面［S］．陕西省食品安全地方标准．
上海健鹰食品科技研究所．2006．凉拌小吃新品——水晶凉皮［J］．农产品加工，(10)：48．
王长江．1997．大米凉皮的制作技术［J］．农家参谋，(04)：30．
吴伟．2014．在家能做的凉皮［J］．农村百事通，(11)：74-75．
周翠英．2010．凉皮、粉皮的制作［J］．乡村科技，(07)：25．

兰州牛肉面

Lanzhou Beef Noodles

Lanzhou beef noodles, also known as "Lanzhou clear soup beef noodles", are "one of the top 10 noodles in China". They are a local snack in lanzhou city, Gansu province. It has been rated as one of the top 3 Chinese fast food by China Cuisine Association, and praised as "the Top Noodles in China". Lanzhou beef noodles are considered the best among all kinds of the noodles in terms of its color, aroma, taste, shape and making method. As the Yellow River runs through the city, Lanzhou is the birthplace of beef noodles. As their history are as long as the Yellow River and the ancient silk road, Lanzhou beef noodles are not only popular in China, but also spread to the whole world.

The authentic Lanzhou beef noodles were created by Ma Baozi in 1915. At that time, due to his poor family condition, Ma Baozi made hot beef noodles at home and sold along the streets of city center. Later, he poured the soup of stewed beef and goat liver into beef noodles, adding the fragrance of noodles. They were known as "hot pot noodles" at that time. In 1919, instead of selling along the street, he opened his own restaurant and served the customers with a bowl of free soup. The waiter immediately served a bowl of fragrant beef soup when the customers entered, which brought Ma Baozi's soup noodles a good reputation. In 1925, his son Ma Jiesan took over the business. Ma Jiesan continued to improve the soup and beef noodles. Later the beef noodles were known nationwide and had the praise of "beef noodles can attract people down from a horse when smelling and stop one's car when knowing the taste". They became one of the famous snacks in Lanzhou with the unique flavor of "clear soup, soft beef and thin noodles" and the characteristics of "one clear (clear soup) , two white (white radish) , three red (red chili oil) , four green (green coriander and green garlic sprout) , five yellow (yellow noodles)", and were widely praised by customers domestically and internationally.

When a bowl of delicious Lanzhou beef noodles is served, you will fix your eyes on the clear and bright soup, white radish, red chili oil, green coriander and garlic sprout, and yellow noodles, and you would like to have a try immediately. This is a nice enjoyment for their color, aroma and fragrance. Fragrant vinegar and chili oil, fresh beef, and soft but chewy noodles will together hit your tongue. Among the bustling crowds in the beef noodles restaurant, the sounds of ordering, ticketing, and eating all intertwined. At the moment, you can experience the unique enthusiasm and boldness of the Northwesters in the restaurant.

"A bowl of beef noodles is everywhere throughout the country." Noodle shapes, soup flavors, type of side dishes, flour brands, consumer preferences, tendency of general taste in different provinces constitute the big data of beef noodles. Lanzhou beef noodles can be found all over the world, and the total number of beef noodles restaurants globally exceeds that of KFC and McDonald's. According to statistics, at present, there are more than 50 000 Lanzhou beef noodles restaurants nationwide, and more than 560 000 people make a living by this business, with an annual turnover of nearly 20 billion yuan. Lanzhou city, the birthplace of beef noodles, has a population of more than 3.6 million and there are more than 1200 beef noodles restaurants and over 10 000 related workers with an annual turnover of 1.5 billion yuan. It has become a highlight in Lanzhou city's GDP growth.

The Legend has it that in the 1940s, there was a "hot pot noodles of Ma Baozi" restaurant in Jiuquan road of Lanzhou city. A bearded official frequently came here to eat. One day, when he asked about the name of the restaurant, the owner answered its name. However, the official said that the name of hot pot noodles sounded a bit weird, and you could call it "clear soup beef noodles" for the noodles' clear soup and delicious taste. The bearded official is Yu Youren. After He returned to Chongqing city, Lanzhou beef noodles became famous throughout the country with his recommendation and praise. From this point of view, Yu Youren is the earliest spokesman for Lanzhou beef noodles.

Lanzhou beef noodles have the characteristics of "one clear (clear soup), two white (white radish), three red (red chili oil), four green (green coriander and green garlic sprout), five yellow (yellow noodles)". Slices of white radish, green coriander and green garlic sprout, and light yellow noodles are all delicious.

兰州牛肉面，又称兰州清汤牛肉面，是“中国十大面条”之一，是甘肃省兰州地区的风味小吃，1999年被确定为中式三大快餐试点推广品种之一，被誉为“中华第一面”。2010年中国烹饪协会将甘肃省兰州市命名为“中国牛肉面之乡”。和黄河、丝绸古道一样悠长的兰州牛肉拉面，不但香飘中国大地，也香飘万里、走向世界。

一 牛肉面的起源及发展

相传兰州牛肉面是马保子始创于1915年的。当时马保子家境贫寒，为生活所迫，他在家里制成了热锅牛肉面，肩挑着在城里沿街叫卖。后来，他又把煮过牛、羊肝的汤兑入牛肉面，奇香扑鼻，当时称为“热锅子面”。1919年他开了自己的店，不用沿街叫卖了，推出了免费的“进店一碗汤”，客人进门来，伙计就马上端上一碗香热的牛肉汤请客人喝，汤爽，醒胃。马保子的清汤牛肉面名气大振。1925年，该店由其子马杰三接管经营，马杰三继续在“清”字上下功夫，不断改进牛肉拉面，直到后来名振各方，被赠予“闻香下马，知味停车”的称誉，成为兰州著名风味小吃，以“汤镜者清，肉烂者香，面细者精”的独特风味和“一清二白三红四绿五黄”的特征，赢得了国内乃至全世界顾客的好评。

2018年4月，《甘肃·兰州—天津合作支持牛肉面产业发展实施方案》（以下简称《方案》）正式印发，《方案》明确省市商务部门联合实施牛肉面提升行动，用3年时间，使兰州牛肉面在天津连锁发展店面数目达到90～100家，每年以30家以上的增幅发展，实现“三年百店”的发展目标；每年安排贫困户300人创业就业，实现“三年千人”行业扶贫计划；同时，进一步带动兰州牛肉面企业实现全国连锁化经营，加快国际化发展的步伐。

二 牛肉面的逸闻趣事

（一）兰州牛肉面赞美诗

清代著名学者张澍在《雨过金城关》中赞美“马家大爷兰州牛肉面”：

雨过金城关，白马激溜回。几度黄河水，临流此路穷。
拉面千丝香，惟独马家爷。美味难再期，回首故乡远。
日出念真经，暮落白塔空。焚香自叹息，只盼牛肉面。
入山非五泉，养心须净空。山静涛声急，瞑思入仙境。

清代王亶望曾作《兰州牛肉面吟》：

兰州拉面天下功，制法来自怀庆府。
汤如甘露面似金，一条入口赛神仙。

（二）兰州牛肉面最早的代言人

传说在20世纪40年代，兰州酒泉路马保子热锅子面馆，一个大胡子官员常常光顾这里。有一次吃饭，他问起店名，老板如实相对。他说，这热锅子面不中听，你看这面汤清肉烂，看着美吃着香，不如叫“清汤牛肉面”。这位老者，就是国民党元老，当时的监察院院长、大书法家于右任。他后来回到重庆，经他揄扬，马保子牛肉面在全国声名鹊起。从这一点来说，于右任先生是兰州牛肉面最早的代言人。

 兰州牛肉面制作工艺

（一）原辅料介绍

兰州牛肉面的制作食材主要有高筋面粉、牛肉、萝卜、蒜苗、香菜、蓬灰、辣椒、香辛料、醋等。

兰州牛肉面之主料——高筋面粉

一般要选择新鲜的高筋面粉，兰州牛肉拉面专用粉。不宜选择陈面，更不宜选择虫蛀、鼠咬、霉变的污染面粉。天然的小麦面粉，并非洁白如雪，而是微微泛黄。高筋面粉是制作牛肉面的首选原料，赋予了牛肉面特有的弹性和嚼感。

兰州牛肉面之配料——牛肉

兰州牛肉面的牛肉主要是产自甘南和青海的牦牛或黄牛。牛肉富含蛋白质，氨基酸组成比猪肉更接近人体需要，有助于提高机体抗病能力。

兰州牛肉面之配料——萝卜

兰州牛肉面精选西北特有的绿头萝卜，其特点为皮厚、肉细、质密、味甘。选用皮色光滑鲜亮、质地坚实无糠心、不霉烂、无虫蛀者为好。

先将萝卜洗净，去其毛根和头尾，切成圆形或扇形的片（半径2.5cm，厚0.2cm），开水焯后凉水冷却，去萝卜的异味；另起一锅，在煮好的牛肉汤中放入萝卜片，煮熟捞出，冷水中漂凉备用。

牛肉汤加煮熟的白萝卜片可祛除牛肉的一部分膻味，增加香味，而且白萝卜有生食开胃、熟食滋补的作用，又可以增加牛肉面的营养。

兰州牛肉面之配料——蒜苗

蒜苗含有辣素，可以对醒脾气、消积食起到一定的作用。优质蒜苗大都叶柔嫩，色鲜绿，叶尖不干枯，不黄不烂，株颗粗壮，整齐、洁净不折断，毛根白色不苦味，而且辣味较浓。蒜苗要洗净、切碎（0.2～0.3cm），切记不可刀剁。兰州牛肉面撒上蒜苗不仅色彩碧绿好看，而且使牛肉面味道更为鲜美。

兰州牛肉面之配料——香菜

香菜中含有很多挥发性物质，其特殊的香气能祛除肉类的腥膻味，起到祛异增香

的作用。香菜洗净、切碎成末，待用。

兰州牛肉面之拉面助剂——蓬灰

蓬灰是用戈壁滩所产的蓬草烧制出来的碱性物质，具有一种特殊的香味，使拉面爽滑透黄、筋道有劲。兰州牛肉面目前使用或曾经使用过的拉面助剂主要有传统蓬灰、速溶蓬灰、精纯蓬灰。精纯蓬灰是以传统蓬灰为原料，采用再结晶法除去传统蓬灰中铅、砷等对人体有害的成分，然后与食盐（NaCl）、面碱（Na_2CO_3）、食用级碳酸钾（K_2CO_3）等混合而成的新型复合拉面助剂。

佐菜

兰州牛肉面通常佐以卤制或酱制的牛腱子肉、卤鸡蛋及各类凉拌小菜（常见如土豆丝、萝卜丝、海带丝、酸菜等）食用。

（二）兰州牛肉面之“灵魂”——清汤制作

牛肉面的精华是牛肉汤。在熬制牛肉汤时，按比例配放干姜片、花椒、小茴香、草果、胡椒、三奈、良姜等几十种调料和中药，充分体现了典型的中国养生文化和中医文化。

牛肉清汤制作工艺流程

选料→浸泡→煮制→撇浮沫→下调料→再煮→捞肉→肉加工→吊汤→调味→成品

参考配方

主料：牦牛肉或黄牛肉 50kg。

配料：牛腿骨 5kg、牛油 2kg、羊肝 1kg、土鸡 1.5kg、绿萝卜、清油、葱花、食盐、香菜、蒜苗、辣子油酌量。

煮肉时调料配方：干姜片 200g，花椒 180g，小茴香 120g，草果 100g，肉桂 100g，胡椒 90g，三奈 50g，肉蔻 50g，良姜 40g，香茅草 40g，荜拨 30g。其中煮肉料与汤的比例为 0.5%～0.7%。

调汤料的配比：干姜粉 90g，花椒粉 85g，胡椒粉 72g，草果粉 55g，桂子粉 32g。可调清香型牛肉面汤 300 碗，每碗牛肉汤 400～500ml。其中调汤料与汤的比例为 0.3%～0.4%，盐与汤的比例为 1.4%～1.5%，鲜姜汁、大蒜汁与汤的比例为 0.1%～0.2%。在调好的汤中加入鲜姜汁和大蒜汁（两者各占 50%）味道更加鲜美，味精与汤的比例为 0.2%～0.4%。

操作要点

1．选料：制汤选用牛腿骨、精牛

肉、羊肝，将牛肉切成2.5kg大小的块，清水清洗。

2．浸泡：以凉水浸泡牛肉3～4h，如有新鲜的牛骨头用斧子砸碎和牛肉一起浸泡。浸泡过的水不可弃去，留作吊汤用。

3．煮制：将浸泡过的牛肉、牛腿骨、土鸡放锅中（不能用铁锅，铁锅易使汤汁变色），冷水下锅，大火烧开，撇去浮沫；将拍松的姜和调料包、精盐下入锅内，用文火煮制，始终保持汤微沸。煮制2～4h后，捞出牛肉、腿骨、土鸡、姜和调料包。如遇腥臊味浓的牛肉，可加入一大厚片生萝卜，煮绵软后捞出，再撇去浮沫。大火煮容易导致牛肉水分蒸发，肉质柴、发干。

4．吊汤：将浸泡牛肉的血水倒入牛肉汤中，使蛋白质受热后产生絮状凝固，在凝固过程中吸附汤中悬浮状颗粒及沉淀物，并上浮在汤面，把这些上浮物撇净后，汤汁澄清。若需要更为鲜纯的汤，则需“二吊汤”或“三吊汤”。

5．调味：一般在临开市前30min调味，将调汤料按比例加入原汤，微火保温。另需备好羊肝汤，将羊肝切小块，放入另一锅里加水1.5kg，煮开撇浮沫。羊肝煮至八分熟加盐慢火煮熟捞出，汤澄清，待调牛肉汤时根据牛肉汤的浓度掺入适量熬好的羊肝汤即可。

煮好的原汤或调好味的牛肉汤装在烧制的陶瓷缸中存放，一般不要超过8h，避免出现发黑或发红、味不正的情况；有些色泽较深的调料最好在煮制时整粒加入，在调汤时尽量使用色泽较浅的调料粉，这样可最大程度避免汤调好后短时间变色。

（三）兰州牛肉面之点睛——辣椒油

参考配方

主料：辣椒面1kg，芝麻100g，熟榨菜籽油5kg，蒜蓉250g。

辅料：大葱150g，圆葱150g，大蒜100g，花椒100g，八角15g，草果、小茴香各50g。

操作要点

选用辣度适中，颜色鲜艳的辣椒面，油选用一级精炼菜籽油。先将油烧热（菜籽油炼去浮沫烧熟），放入葱段、姜片、砸破的草果、小茴香炸出香味，待油温降至120℃左右时，捞出调料。在辣椒面中放少许盐，倒入温油炸，加入芝麻搅拌至均匀晾凉，下蒜蓉即可。一般500g辣椒面用2500～3500g油，炸透后放置24h以后备用。也可以根据下料的顺序，油温逐渐下降，形成明显的分层。160℃第一次下辣椒面，部分辣椒面焦煳，突出煳辣香味；140℃第二次下辣椒面，突出香辣味；小于120℃，第三次下辣椒面，油温低香味不足，但颜色好。

(四)面的制作

兰州牛肉面制作工艺流程

选面 → 和面 → 饧面 → 溜条 → 拉面 → 煮面 → 成品

操作要点

1．选面：一般要选择新鲜的高筋面粉，不宜选择陈面，更不宜选择虫蛀、鼠咬、霉变的污染面粉。新鲜的高筋质面粉，蛋白质含量高，是拉面制作成功的保证。蛋白质含量在12%以上的面粉都可以作为兰州牛肉拉面的使用面。

2．和面：和面是拉面制作的基础，俗有“七分和面三分拉”的说法。

配比：面粉500g，食盐4g，拉面助剂8～10g，水250～300g。

水温：和面时水的温度一般要求冬天用温水（18℃左右），其他季节则用凉水。和好的面团温度保持在26～30℃，面粉中的蛋白质吸水性最高，可以达到150%，面筋的生成率也最高，质量最好，即延伸性和弹性最好，最适宜抻拉。若温度低于26℃，则蛋白质的吸水性和质量会随温度的下降而下降；超过30℃，同样也会降低面筋的生成；当温度到60℃时候，则会引起蛋白质的变性，而失去其性能。

加水方法：分三次加水。和面时，将所用面粉倒在案板上，在面粉中挖坑，水分三次加入，第一次先用60%～70%的水把面团用劲翻拌、搓揉，调成“梭状”，拌成梭状是为防止出现包水面（即水在大面团层中积滞），因包水面的水相和粉相分离，致使面团失去光泽和韧性，影响面条拉制过程和熟后的口感；第二次撒20%的水翻拌、搓揉均匀；第三次根据面的软硬程度，再撒10%的水搓揉至面团滋润。面团如不滋润，可适量蘸水再揉，揉好的面团用塑料纸盖严以防干裂。在使用时，加蓬灰水揉匀使面团松软能拉开即可。

加水量：面与水的比例，根据面粉的质量掌握，一般500g面粉用水250～300g。

和面方法：和面时采用捣、揣、登、揉等手法，“捣”是用手掌或拳撞压面团；“揣”是用掌或拳交叉捣压面团；“登”是用手握成虎爪形，抓上面团向前推捣；“揉”是用手来回搓或擦，把面调和成团。和面主要就是需要捣面，双拳（同时沾蓬灰水，但要注意把水完全打到面里）击打面团，非常关键的是当面团打扁后再将面叠合时一定要朝着一个方向（顺时针或逆时针），否则面筋容易紊乱，此过程大约得15min以上，一直揉到不沾手、不沾案板，面团表面光滑为止。有一个非常简单的小现象，把面揉到起小

泡泡了就差不多了。捣、揣、登是防止出现包渣面（即面团中有干粉粒），促使面筋较多地吸收水分，充分形成面筋网络，从而产生较好的延伸性。

3．饧面：将揉好的面团表面刷油盖上湿布或者塑料布，以免风吹后发生面团表面干燥或结皮现象，静置 30min 以上，使面团中央未吸足水分的粉粒充分吸水，防止面团产生小硬粒或小碎片，使面团均匀，更加柔软，并能更好地形成面筋网络，提高面的弹性和光滑度，制出成品也更加爽口筋道。

4．溜条：揉面时两脚稍分开，站成丁字步形，上身弯曲成角，身体不靠近案板，这样使劲用力揉面时，才不致推动案板，其手法有揉、扎、擦三种。揉面的同时，把面往开揉搓，双手交替，一手接一手排过去，要注意排均匀，将面揉开推平在案板上，加适量的蓬灰水，再用力搋（chuāi）捣，这样连续重复数次，至手擀面团有弹性滋润为止（以面不沾手或能拉开为准）。牛肉面讲究“三遍水，三遍灰，九九八十一遍揉”。

溜条即在大团软面反复捣、揉、抻、摔后，将面团放在面板上，用两手握住条的两端，抬起在案板上用力摔打。条拉长后，两端对折，继续握住两端摔打，业内称其为顺筋。然后搓成长条，抹上少许菜籽油，揪成 30mm 粗、筷子长的一条条面节，或搓成圆条。盖上油布，饧 5min 左右，即可拉面。

5．拉面：将溜好的面条放在案板上，撒上清油（以防止面条粘连），然后随食客的爱好，拉出形状和粗细不同的面条。

拉面时，手握两端，两臂均匀用力加速向外抻拉，然后两头对折，两头同时放在一只手的指缝内（一般用左手），另一只手的中指朝下勾住另一端，手心上翻，使面条形成绞索状，同时两手往两边抻拉。面条拉长后，再把右手勾住的一端套在左手指上，右手继续勾住另一端抻拉。抻拉时速度要快，用力要均匀，如此反复，每次对折称为一扣。一个面节拉一大碗面，每拉一下，要在手腕上回折一次，拉到最后，双手上下抖动几次，则面条柔韧绵长，粗细均匀。

根据兰州牛肉面标准，毛细、细面、二细面直径分别为 0.1cm、0.2cm、0.3cm，粗细均匀，不粘连，不断条；韭叶、宽面、大宽面直径分别为 0.5cm、1.5cm、2.5cm，厚薄均匀，宽窄一致，不粘连，不断条；荞麦棱粗细均匀，棱角分明，不粘连，不断条。

6．煮面：将拉好的面放入沸腾的面锅里，等浮起来后用竹筷将粘在一起的面条拨开，再翻两遍即可出锅，不可煮得时间太长（30s～1min），否则使面条太烂容易粘连在一起，口感不好，且失去牛肉面的香味。有句顺口溜形容往锅里下面：“拉面好似一盘线，下到锅内悠悠转，捞到碗里菊花瓣”。

面熟后捞入碗内，放入牛肉汤、萝卜、肉丁（片）适量即成，并以每个人的口味加上适量的香菜、蒜苗、葱花及辣子油。按每碗中熟面条净含量分为大碗兰州牛肉拉面（不得少于 275g）和小碗兰州牛肉拉面（不得少于 175g）两种规格；每碗面中肉汤净含量 300～500g，熟牛肉 30～50g。

四 牛肉面的风味特色

兰州清汤牛肉面有“一清、二白、三红、四绿、五黄”五大特点，即牛肉汤色清气香，萝卜片纯净洁白，香菜、蒜苗青嫩翠绿，红油辣子鲜香诱人，面柔滑透黄，劲道味足，满口留香。

兰州牛肉面营养均衡合理，富含蛋白质的高原牦牛肉与高原地带产的优质面粉有机结合。牦牛肉滋养脾胃，强筋健骨。辅以羊肝和牛腿骨熬制的肉汤，汤香浓郁。羊肝含多种维生素及微量元素；牛腿骨富含钙、磷等营养成分。汤中加入的数十味调料，除了使汤香味浓外，各种调料都兼有对人体有益的功用。汤中的白萝卜去除肉腥，更能起到补中益气、帮助消化的作用。红辣椒弥补了牛肉及面条中维生素C含量不足的缺陷。面条中使用的蓬灰水呈碱性，可中和肉汤中的酸性，并使面条筋劲黄亮，是兰州牛肉面独有的特色。

参考文献

陈小丽. 2012. 兰州牛肉面品牌的文化拓展问题探析 [J]. 社科纵横，(07)：125-127.
董晓君. 2017. 舌尖上的文化共享：兰州牛肉面的前世今生. [J]. 今日民族，(05)：48-51.
段广亭. 2008. “金城”美食八宝——不容错过的饮食文化 [J]. 农产品加工，(11)：42-43.
蒋凌. 2016. 兰州牛肉面更待提升“含金量” [N]. 兰州日报，2016-10-26 (013).
李岩. 2015. 兰州牛肉面拉面剂作用机理及安全性综合研究. 食品工程 [J]. (03)：53-56.
柳佳. 2011. 兰州牛肉面产业化经营调查分析 [J]. 合作经济与科技，(16)：84-86.
卢玉清. 2018. 兰州人的集体记忆和文化认同—— 一碗牛肉面 [J]. 传播力研究，(07)：20-21.
马琦明. 2017. 兰州牛肉面文化的传承与发展 [J]. 发展，(12)：44-48.
马竹书. 2012. 兰州牛肉面产业化发展再思考 [J]. 甘肃科技纵横，41 (05)：127-128，40.
牛正寰. 2003. 正宗兰州牛肉面 [J]. 西部大开发，(04)：75.
邱兆蕾. 2015. 浅谈兰州牛肉面的发展现状、问题及对策——从成功企业“金鼎兰州第一面”谈起. 北方经贸 [J]，(11)：55-57.
宋鹏，张雨婷. 2018. 兰州牛肉面商誉的地理标志保护分析 [J]. 西部法学评论，(02)：119-125.
吴季康. 2002. 兰州牛肉面轶事. 西部人 [J]. (02)：35-38.
徐晓兵. 2015. 牛肉面产业化亟需落实战略规划 [N]. 兰州日报，(03).
朱则，薛泽林，张劲松. 2013. 民族文化视角下兰州拉面发展现状调研 [J]. 民族论坛，(09)：100-104.

新疆烤馕

Xinjiang Baked Naan

As one of the main flour-made foods of all ethnic groups in Xinjiang, Naan boasts a history of more than two thousand years. At almost every town in Xinjiang, you can find small workshops for baking and selling Naan (commonly known as Naan house). Naan is easy to be stored for its less water content. It is made from exquisite ingredients. Matched with milk tea or broth, it tastes more delicious.

As an "ecological food" that adapts to natural conditions and the needs of human society, it originated in the southern part of Xinjiang. It was invented by the indigenous farming peoples in Xinjiang oasis in light of the local dry climatic conditions with a lack of rainfall. During the development of Uyghur in the southern part of Xinjiang and the evolution of the Uyghurethnic group, this ecological food has been passed down and gradually developed, eventually forming a Xinjiang food culture of strong regional characteristics with Naan at its core. There is no unified statement about the origin of Naan, mainly including the introduction from Western and Central Asia, an invention by the Uygur, and the introduction from the Central Plains.

There is a long history of the Naan, and archaeologists have unearthed ancient Naan in ancient tombs. According to research, the word " 馕 " is derived from Persian and is circulated in the Arabian Peninsula, Turkey, and Central Asian countries. The Uyghur originally called it "Aimaike", and it was not until the introduction of Islam into Xinjiang that it was called "Nang". Legend had it that when Monk Tang went to the west across the Gobi desert to fetch the scriptures, the food that he brought with him was Naan, which helped him to complete the journey full of hardships.

Many famous poets in Chinese history have also described Naan in their poems. Bai Juyi said in his poem *To Hubing and Yang Wanzhou*: "The capital sets an example for the cooking of Hubing. The crisp pancake is newly fried by fragrant oil. Give it to the hunger-stricken Yang Wanzhou, see if it is to his liking."

The ancient proverb "One can live without dishes a day, but cannot live without Naan" expresses the important position of Naan in the lives of the Uyghur people. On some occasions, Naan also shows a special meaning. For example, the Naan residues that fell on the ground should be picked up and placed high for the birds to eat. The Uyghur people also regard baking Naan as a symbol of mascots and happiness. For example, when a man proposes to a woman, clothes, salt, box sugar, and especially five Naans are sent as the gift. At the wedding ceremony, there should be a girl holding a tray with both hands on which a bowl of salt water with two small Naans was put. The girl will stand between the bride and groom, letting them rush to eat these two pieces of Naan in saltwater, which symbolizes harmony and everlasting love. At this point, the bride and groom rushed to grab the Naan in the bowl. The person first catches the Naan is more loyal to love. "Grab the Naan when the time comes." Naan grabbing could become the first climax of the wedding.

On December, 2017, "Uncle of Naan" from Nilek county in Yili prefecture became a hit. An artistic "Golden Chicken Naan" he baked sold tens of thousands of Yuan.With the support of the local government, "Uncle of Naan" founded the "Naan factory" and sold 8 000 pieces of Naan to the local, Urumqi, Karamay and Shenzhen, and helped nearly 40 families to get rid of poverty. With the help of the Subutai Township Government, he established the Beinamu Naan Cooperative in Subutai Township. In order to improve the popularity of the "Naan Factory", he creatively baked such artistic Naan as "Naan of Heavenly Horse", "Naan of Golden Chicken" and "Naan of Bottle Gourd", and brought them to many places for finding a market.

The Naan is golden yellow on the surface with red-brown color in its edges. It has a thin crust with the skin and fillings mixed evenly. Layer by layer, it is neither too soft nor too hard. Its shape is complete. It is sweet, crisp, delicious, tender, fresh but not greasy. It has a refreshing fragrance and delicious flavor. Naan has such features as a unique cooking method, different varieties, low water content, and abundant nutrition.

馕是新疆各族人民喜爱的主要面食之一，已有两千多年的历史。走遍新疆的每个城镇几乎都能找到烤馕、卖馕的小作坊（俗称馕房）。馕含水分少，易于储藏；用料讲究，富有营养，配以奶茶或肉汤，吃起来更是香酥可口。

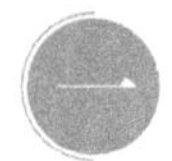

馕的起源及发展

（一）馕的起源

馕是农耕土著民族发明创造并适应当地干旱少雨气候条件的产物，是适应自然条件和人类社会需要的“生态食物”。在新疆南部回鹘化、伊斯兰化的过程中以及维吾尔族的演变过程中，馕这一生态食物得到传承并逐步发扬光大，最终形成以馕为核心的、具有很强地域特色的新疆饮食文化。

馕的起源没有统一的说法，主要有维吾尔族发明说、中亚和西亚传入说、中原传入说等。

维吾尔族发明说

馕起源于新疆，是维吾尔民族创造的。这是现今大部分人的观点和认识。如奇曼·乃吉米丁、热依拉·买买提在《维吾尔族饮食文化与生态环境》一文中认为：“烤馕在维吾尔族饮食文化中已有悠久的历史，追溯源头，可以从维吾尔族祖先回鹘算起，烤制品是游牧民族的主要食品。”何婧云在《维吾尔族“馕”文化及其当代转型》一文也提出：“维吾尔人烤馕的历史距今已有 2000 余年，最初他们称其为俄克买克，称之为馕，是从伊斯兰教传入新疆后开始的。”

中亚和西亚传入说

馕是从中亚和西亚传入我国的，它是农耕文明的结果。新疆社会科学院的夏雷鸣在《西域薄馕的考古遗存及其文化意义》一文中也提出，一种源自波斯叫“恰帕提”的“薄馕”，“作为波斯饮食文化的使者，向西经丝绸之路，传到了阿拉伯国家，向东翻越葱岭传到了新疆”。

中原传入说

馕来自中原地区，是由中原传入西域的。持这种意见的人认为，南阳附近有一种当地人俗称“ganglou”，方志上写作“缸炉”的饼子，而“缸炉”应写作“干锣”。“干锣”就是用北方最常吃的麦面做成的饼子。“干锣”和新疆的馕一模一样，而新疆馕的发音又恰恰跟南阳方言里“好”的发音一样。南阳在汉朝时为东西交流重镇，“干锣”作为远行的干粮佳品就会自然而然地流入西域。传往西域后当地人就把“干锣”美食称为“nang”，写成汉字就是馕。

（二）馕的发展

馕的发展历史悠久，考古学者曾在古墓出土过古代的馕。据考证，“馕”字源于波斯语，流传在阿拉伯半岛、土耳其、中亚和西亚各国。传说当年唐僧取经穿越沙漠戈壁时，身边带的食品便是馕，是馕帮助他走完充满艰辛的旅途。

新疆维吾尔自治区博物馆陈列有吐鲁番出土的唐朝的馕，说明在一千多年前，吐鲁番人就会做馕了。馕古代称“胡饼”“炉饼”。贾思勰著《齐民要术》摘录的《食经》中有关于做馕的技术资料，可见馕在中国食谱中由来已久。

现代馕的制作仍在不断发展。2017 年 12 月，伊犁州尼勒克县的“烤馕大叔”火了！他烤制的一个具有艺术性的“金鸡馕”卖出了上万元的价钱。“烤馕大叔”在当地政府的支持下，成立了苏布台乡贝纳木馕合作社。合作社里并排立着 15 个馕坑，每天烤制上万个馕，被当地人称为“馕工厂”。为了提高“馕工厂”的知名度，他创意烤制出“天马馕”“金鸡馕”“葫芦馕”等造型的艺术馕，每日向当地、乌鲁木齐、克拉玛依以及深圳等地销售 8000 个馕，带领近 40 个家庭脱贫致富。

馕的逸闻趣事

中国历史上许多著名诗人在他们的诗篇中还描写过馕。如白居易的《寄胡饼与杨万州》：“胡麻饼样学京都，面脆油香出新炉。寄予饥馋杨大使，尝看得以辅兴无。”

“宁可一日无菜，不可一日无馕”的古老谚语表达了馕在维吾尔族人民生活中的重要地位。在一些场合里，馕还表达着特殊的含义。维吾尔人会把烤馕看作吉祥物和幸福的象征，比如男方向女方提亲，作为见面的礼物除了衣料、盐、方块糖，还必须有五个馕。在结婚仪式上，要安排一位姑娘，双手捧出一个托盘，上面放着一碗盐水，盐水里泡着两块小馕。姑娘就站在新郎新娘中间，让他俩抢着吃下这两块象征着爱融洽、有福同享、白头偕老的盐水馕。此时，新郎新娘争先下手去捞碗里的馕，谁先捞到馕，就表示谁最忠于爱情。“该出手时就出手”，抢馕就成为婚礼中的第一个高潮。

馕的种类

馕的种类非常丰富，分类各异。

根据用料分类

传统的馕一般呈圆形，大致可以分为以下几种。

1．白馕：白馕主要用小麦粉以发酵或不发酵的方法制作而成，主要包括阿克馕（葱花馕、片馕）、昆具提鲁克馕（芝麻馕）、给日德馕（窝窝馕）、西克热馕（甜馕）、希尔曼馕、艾买克馕（大馕、薄馕、大薄馕）、考买其馕（灰烬馕）。

2．蒸馕：蒸馕又称奥尔馕。盐水和小麦粉发酵后，分成若干大小相等的剂子并擀成薄片，抹上奶皮子和切碎的奶酪、洋葱末、胡萝卜丝、羊肉末、香菜等进行调色、调味，然后卷起来放在笼里蒸制而成。

3．粗粮馕：粗粮馕主要是以玉米面为主，或在小麦粉中掺杂一定量的玉米面、大麦面或是高粱面发酵制成的馕，有居万德馕（混合面馕）、阔休克卢克馕、扎合热馕（玉米馕）、苞谷馕、喀瓦馕（南瓜馕）。

4．油馕：油馕是以小麦粉和植物油为主料制成的含油量较高的馕，主要有托喀奇馕（油馕）、素特桃喀奇馕（奶子馕）、卡特里玛馕（千层馕）。

5．肉馕：肉馕以小麦粉制作面坯，以羊肉和洋葱为主料制作馅料，具有皮脆、肉多、油肥、味香的特点，有古西馕和古西给日德馕。

6．果仁馕：果仁馕以葱花馕、芝麻馕、大馕、甜馕面坯制作工艺，制成中间厚四周薄的圆形或花型馕坯，烤制前在馕坯表面沾满或嵌满果仁的馕，常见的有瓜子馕、核桃馕、花生馕等。

7．其他馕：以小麦粉为主，并充分利用新疆地产的优质果仁（如核桃仁、巴旦木、杏仁）、果脯、果酱、牛羊肉、植物油、羊油、牛奶、鸡蛋等资源，改进了馕制作的配方和工艺，研制出的现代式改良品牌馕。

根据形状分类

从馕的形状上来看，传统的馕为圆形馕，也有区别于传统烤馕的艺术烤馕，有长方形、正方形、菱形、塔形、哈密瓜形、香蕉形、树叶形、葡萄形等。其中最大的馕叫“艾曼克”馕，中间薄，边沿略厚，中央戳有许多花纹，直径足有40～50cm。这种馕大的需要1～2kg面粉，被称为馕中之王。最小的馕和一般的茶杯口那么大，叫“托喀西”馕，厚1cm多，是做工最精细的一种小馕。还有一种直径约10cm，厚5～6cm，中间有一个洞的“格吉德”馕，这是所有馕中最厚的一种。

四 馕的制作工艺

（一）主要原料及辅料介绍

馕的品种很多，但古今制馕的主要原料和基本方法没有发生太大变化。它的主要原辅料是：面粉（小麦粉或玉米面粉）、芝麻、洋葱、鸡蛋、清油、牛奶、糖、盐。

馕的主料——高筋小麦粉

高筋小麦粉含有丰富的碳水化合物和营养物质。选择面粉时可以按照看、闻、选的步骤进行选择。选用具有麦香味、自然乳白色或略带淡黄色的面粉。

馕的配料——羊肉

羊肉含有丰富的优质蛋白、钙、铁、维生素D。选色泽淡红、肌肉发散，肌纤维较细短，肉不粘手，质地坚实，脂肪呈白色或微黄色，质地硬而脆的羊肉。

馕的配料——洋葱

洋葱含有较少的热量，富含钾、维生素C、叶酸、锌、硒，及纤维质等营养素，且洋葱不含脂肪，含有可降低胆固醇的含硫化合物；其气味辛辣，能除去腥味、油腻厚味及菜肴中的异味，并产生特殊香气；可刺激胃、肠及消化腺分泌，增进食欲、促进消化。挑选洋葱时，选择色泽鲜亮、水分充足、闻起来气味辛辣、掂起来沉甸甸的洋葱。

馕的配料——植物油

应选择知名品牌的保质期内的植物油。外观上要求色泽金黄、透明，无杂质，闻起来清香扑鼻，无异味。

（二）馕坑的制作

馕坑由耐火土、碱土、土块砖制成。一般馕坑用羊毛和黏土砌筑成，高约1m，肚大口小，形似倒扣的宽肚大水缸，底部架火，底部留有通气口，通常是夯土结构，馕坑四周用土坯垒成方形土台，以便烤馕人在上面操作。新疆南部多选用硝土做馕坑坯，乌鲁木齐则用砖块做馕坑坯。其制作方法如下。

1．选土：戈壁滩上找碱土，好的碱土一般很大，块状，色偏白且不均，内部呈细小不均匀的蜂窝状，有少许粉末，干燥很硬（很硬说明含盐量高，做出来的馕坑经高温和凉水也不会损坏），需要用榔头锤碎，然后压成粉末状。

2．造型：先用砖块垒成一个大致的馕坑，为一个长方体（高0.5m，长宽一般为2～3m）。

3．掏：掏空中间部分，长方体顶部砌出一个和自行车轮一样大的口，底部为直径1～1.5m的圆。长方体内部大约有一个1/4的球体部分是空的，在长方体与地面的接壤处随机开一个5cm×10cm的开口用于通风。

4．涂抹：把粉末状的碱土加水混合至稀泥状，然后涂抹在垒成的砖馕坑上，厚度为5～8cm，整个馕坑都涂抹上，最后风干即可。烤馕前，先将干柴（木炭或煤炭）放在坑底燃烧，待明火消失时，坑壁已烧得滚烫，即可把擀好的馕面坯贴在坑壁上，10min便能烤熟。

（三）普通馕的制作

普通馕制作工艺流程

操作要点

1．发面：发面的时间和烤馕的成功与否有直接关系。一年四季发面的时间不一样，冬季时间较长，一般要5～6h，室内还要加温，保持

发面 → 揉面 → 成形 → 贴馕 → 扒馕 → 成品

在20℃左右；夏季发面时间短，一般2h为宜。无论是冬季还是夏季，还需要通过添加的发酵面团的量来调节发面的时间。发酵面团兑入得多，发酵的越快，兑入少，发酵就越慢。

2．揉面：馕品质的好坏除了与发面时间长短有关以外，还依赖于揉面的功夫。在揉面时讲究揉匀、揉透，馕吃起来才有劲，也不会松散。从发面到下一步的成形（馕坯），尽量多次揉面，揉的次数越多，馕的味道越好，保存时间越长。

3．成形（做馕坯）：揉面时可以根据需求做不同形状的馕坯。可以做成圆凳子面大小的薄馕坯，也可以做成碗口大的厚馕坯。薄馕坯有不同的尺寸，大馕（直径为30～50cm）用来招待客人，而旅游和离家较长时间时携带小馕（直径为8～10cm）。成形后在馕坯表面还要用多种辅料点缀花纹，比如葡萄干、核桃、花生、豌豆、蛋清、芝麻等。铺料不但可提高馕的可食性和营养性，而且会使馕既好吃又好看。

4．调节馕坑温度：发面时应同时对馕坑进行加热，加热要持续一定时间，达到馕坑内表面热量均匀分布。虽然最近在市场上出现的燃气和电加热馕坑不需要加热到很高的温度，但还是要先把馕坑加热到比烤馕时温度略高一点（10～20℃）的温度，保证馕坑表面热量均匀。馕坯的形状不同所需要的温度有些差别，一般烤馕温度在220～230℃。当馕坯帖到馕坑内表面之前，需对馕坑的内部温度进行调节。撒盐水不但可调节馕坑的温度，而且还增加了馕坯与馕坑壁的粘贴度，防止馕从馕坑壁上脱落。同时还可以利用出风口和进风口调节馕坑的温度，使馕坑保持相同的温度。

5．贴馕：馕坑温度调节好后，把馕坯一个一个贴到坑内。贴馕时一般从馕坑的底部贴到坑口，最后盖好坑盖。中间还需要打开坑盖观察馕坯表面颜色的变化，如果热量不够或热量过高，都可以通过馕坑底部的通风口调节热量。馕坯在适当的温度下

烤 10～20min 就能烤熟。

6．扒馕（出坑）：尽管在坑内馕坯离馕坑中心的距离不同，但几乎可在同一时间烤熟。这是因为前面通过调节馕坑温度，馕坑内部达到热平衡状态，馕坑内部各处的温度基本相同。扒馕时用铁铲子把馕从坑内一个一个铲出来，或者用铁碗一次扒几个馕。有时候因馕坑温度调节不合理，可能贴在馕坑某些区域的馕没完全熟。这时先把这种馕扒出来，然后再竖着放到馕坑底部，几分钟后拿出来即可。

（四）肉馕的制作

肉馕的做法有两种：一种是在馕坑里烤；另一种是在油锅里炸。无论哪种做法，其原料大都一致。面和好后要稍发酵。馅主要用羊肉、洋葱、盐和胡椒粉等。

面坯及馅料配方

面坯：面粉 1000g，水 500g，食用盐 10g，干酵母 4g，奶油 50g。

馅料主料：羊肉（肥瘦相间）400g、洋葱 200g。

馅料调料：食盐 8g、味精 4g、孜然 6g、黑胡椒 4g。

操作要点

1．面坯制作：将面粉、干酵母、奶油、食用盐、水混合，搅拌均匀，让蛋白质充分吸水形成面筋网络结构，揉匀制成面团并搓条，将其揪成剂子待用；每个剂子 150g，揪好的剂子揉圆，盖薄膜饧制 3～5min。

2．馅料制作：选择新鲜肥瘦相间羊肉，羊肉切 0.5cm 小丁，洋葱切 0.5cm 小丁，加盐、孜然、味精和黑胡椒碎拌匀成馅。

3．包馅：将松弛好的面坯压扁，擀制成 20cm×12cm 的长方形面皮，将调制好的肉馅均匀平铺在擀制好的面皮上，四周留 1cm 左右的空白，铺好之后，从长的一边开始，卷成长棍状，卷好之后再饧制 3min，搓长至 80cm，压住一头盘起成圆饼状，上面撒少许面粉，用手掌压扁为直径 18cm 大小的饼状。

4．焙烤：馕坑烧热，向馕坑内壁撒盐水，待水分蒸发，盐粒析出，将肉馕生面坯贴入馕坑壁，温度控制在 220℃，烤 20min 左右，至馕面呈金黄色时取出即可。

肉馕不仅保持了馕外表金黄、光亮、酥脆的特点，而且改善了馕的品质和风味，丰富了馕的品种。

五 馕的风味特色

馕表面金黄色，色泽均匀，四周下边为棕红色；皮薄，软硬适中；外形完整；香甜脆嫩，鲜香不腻，清香宜人、可口，无异味。具有工艺独特，种类繁多；水分含量少，易于保藏；营养丰富等特点。

从馕的制作过程看，它很好地达到了热与寒、环境与人的平衡。从制作馕的基本原料看，大多为粗粮，性凉，对润通肠道、保养脾胃、调节气血有帮助。馕的佐料芝麻、洋葱、孜然、黑胡椒等，性温，可健脾胃，祛寒痰，有一定的活血、利尿、杀菌、降血脂等作用；洋葱对预防高血压有帮助。

参考文献

阿布都艾则孜·阿布来提，艾力·如苏力，博尔汗·沙来，等. 2012. 维吾尔族馕坑几何结构及新能源馕坑的设计［J］. 食品科技，(11)：79-82.

阿布都艾则孜·阿布来提，博尔汗·沙来，阿拜，等. 2012. 黑体辐射应用——馕坑工作原理及最佳工作状态的实验和理论研究［J］. 食品科技，(01)：140-143.

阿布都艾则孜·阿布来提，买买提热夏提·买买提，艾力·如苏力，等. 2015. 馕的加工工艺与烤馕机的工作原理［J］. 安徽农业科学，(18)：286-288.

阿尔斯朗，阿孜古丽·依明，卢秀莲. 2004. 新疆维吾尔族主食“馕”的营养及养生作用初探［J］. 中国民族医药杂志，10(03)：43-44.

阿依夏木古丽·阿尤甫，苏比努尔·阿里木. 2016. 维吾尔族馕文化资源及其开发价值分析［J］. 中国民族博览，(12)：60-61.

艾麦提·巴热提，热合满·艾拉. 新疆馕储藏保鲜研究［J］. 现代食品，1(02)：48-52.

安尼瓦尔. 哈斯木. 2017. 馕·馕坑与馕文化漫谈［J］. 新疆地方志，(02)：53-58.

曹雪琴，姜林慧，王文文，等. 2019. 新疆特色食品馕的营养成分分析［J］. 营养学报，41(01)：99-101.

谷亚文，肉孜·阿木提，史勇，等. 2018. 我国烤馕装置现状与分析［J］. 新疆农机化，(02)：29-31.

何婧云. 2006. 维吾尔族“馕”文化及其当代转型［J］. 农业考古，(04)：256-259，296.

李冬梅. 2000. 浅谈维吾尔族饮食民俗中的文化质点——馕［J］. 西北民族学院学报（哲学社会科学版），(03)：61-67.

李正元. 2012. 馕的起源［J］. 中国边疆史地研究，22(01)：112-117.

马晓瑞. 2018. 也谈维吾尔族馕文化［J］. 安徽文学（下半月），(11)：193-194.

美合日古丽. 艾木杜力，曹竑. 2009. 维吾尔族传统小吃——肉馕生产技术［J］. 西北民族大学学报（自然科学版），30(02)：86-90.

奇曼·乃吉米丁，热依拉·买买提. 2003. 维吾尔族饮食文化与生态环境［J］. 西北民族研究，(02)：155-165.

热莎拉提·玉苏普. 2007. 新疆维吾尔族的特色食品——馕的基本解读［J］. 康定民族师范高等专科

学校学报，16（02）：17-20.
热莎拉提·玉苏普. 2007. 新疆维吾尔族的特色食品——馕的基本解读［J］. 康定民族师范高等专科学校学报，16（02）：17-20.
孙含，王晶，赵晓燕，等. 2018. 新疆特色面制品馕的研究进展［J］. 粮油食品科技，26（06）：19-24.
俞雅琼，张金龙. 2017. 肉馕烤制的标准化制作工艺的研究［J］. 新疆职业大学学报，25（02）：80-82.
夏雷鸣. 2005. 西域薄馕的考古遗存及其文化意义：兼谈波斯饮食文化对我国食俗的影响［J］. 新疆大学学报（哲学·人文社会科学版），33（1）：94-100.
朱晓莹，吐汗姑丽，努尔古丽·热合曼. 2015. 新疆传统烤馕酵子中乳酸菌的分子生物学鉴定［J］. 新疆师范大学学报（自然科学版），34（02）：19-23.

印度飞饼

Indian Flying Cake

"Indian flying cake", also known as "Indian pancake" in Southeast Asia. It is made from a properly sized dough kneaded from high-quality flour, and the dough is spread larger by "flying" it with hands in the air. The freshly made "Indian flying cake"can be thin, crisp, soft, and fragrant. It has golden color with multiple layers. It has crispy crust, tender inside, soft and delicious taste. It is cooked on-site and the whole process deserves appreciation. It is a very popular specialty in India and Southeast Asia.

"Indian flying cake" is known as "paratha roti" in India. It originates from the mountainous areas around the Bay of Bengal, India. The local residents usually make paratha roti with flour of high gluten, coconut juice, butter and condensed milk.

The traditional practice in India is to knead the dough into a small round piece and then to "fly it to thin" the small dough after a few strokes. "Flying" means that holding the edge of the small dough and throwing it back and forth for making it thin by the force of gravity. At the end of the "flying" process, the dough becomes as thin as a piece of paper, which can be seen through from one side to another. After "flying" thin, put it into a small pan and fry it, while fold the paper-thin dough into square or round shape during the frying. There is also a simple and small stove next to the pan, but there is no pot above the flame. When the pancake in the pan is cooked well with a little inflation, the cook will quickly pick it up with his hands and throw it into the fire of the stove next to the pan. Surprisingly, the pancake becomes expanding once it is on the flame as if it miraculously suspends above the flame. Soon, the cook takes it out and puts it on the plate, then the aroma is overflowing around.

As the Indians moved to Malaysia, Singapore and other places, they also brought their food culture to these countries and regions, including "Indian flying cake". The original Indian food retains its basic making methods, meanwhile the taste has been improved to meet the needs of local residents. The "Indian flying cake" in Southeast Asia is generally considered to be a variant of paratha roti in India. For example, it is called roti canai in Malaysia, roti prata in Singapore, roti cane in Indonesia, and ro tee in Thailand.

In Malaysia, Indian flying cake is a round flat or square pancake, and is one of the most popular food at mamak stalls in Malaysia (Malay: Gerai Mamak) . The so-called mamak actually refers to the local muslim Indian residents. Traditional mamak stalls mean eateries operated by local muslim Indians along the roadside. Indian flying cake can be eaten for breakfast, lunch, afternoon tea and dinner, or even night snack. When people eat Indian flying cake, it is common to eat it with several kinds of dal sauce (curry sauce) or sambal sauce. In Malay, Indian flying cake or Indian pancake is called roti canai. The word "roti" means bread or flatbread and it can be understood as the pancake. Meanwhile, "canai" means spreading the dough or squeezing the dough back and forth because the Indian flying cake cooks will squeeze the dough, and "fly" the dough to make it wide and thin.

Legend has it that a long time ago, only a few people could afford white flour in India. Most of the common people mainly ate roughage, which were made from a kind of herbal tubers. One day, a distinguished guest visited a family whose hostess was kind and beautiful, but there was only a small amount of white flour left in the house. What kinds of food should be prepared for the guest? The beautiful and kind hostess used her quick wits and simply blended the roughage flour with the white flour, and prepared baked pancake for the guest. The hostess thought that the roughage flour had a bad taste. If the pancake was too thick, it was not easy to bite. So she made the pancake as thin as a piece of paper and folded it. After eating the baked pancake, the guest was full of praise for the food. When the host tried it, he found that it was delicious for its fragrant and chewy taste. Especially when it was hot, it was crispy and crunchy. Later, the pancake was widely spread in India and improved and brought by the Indians to Southeast Asia and the rest of the world.

Indian flying cake is bright, golden, and semi-transparent with a crisp, mellow, and delicious taste. It can be generally divided into two layers; the outer layer is golden yellow and crisp, while the inner layer is soft and white with a taste of slight sweetness. After it is chewed, its fragrance still lingers inside the mouth.

"印度飞饼"，在东南亚又称为"印度煎饼"，是用调和好的优质面粉揉成大小合适的面团，并通过在空中用"飞"的绝技展开面团。刚做好的"印度飞饼"具有薄、脆、软、香的特点，外表色泽金黄，层次丰富，吃起来外酥里嫩、松软香脆、味道可口，现场制作，观赏性极强，是印度以及东南亚各国的特色风味美食。

一 印度飞饼的起源及发展

"印度飞饼"在印度更为人们熟悉的名字是"帕拉塔"（paratha roti），最早来源于印度新德里孟加拉湾大山脉地区，当地居民常年以面粉、椰浆、黄油、炼乳等制作"帕拉塔"。

印度飞饼传统做法是将面团捏成一个小圆团，擀几下后再"飞"薄。"飞"是指在做饼时，拽住面饼的边缘来回抛动，利用离心力把面饼抛开、抛薄。"飞"到最后，饼变得薄如纸片，透过饼可以看到后面的东西。"飞"薄后把饼放入小平底锅中煎熟，同时将面饼折叠成方形或者圆形。当平底锅中的饼快熟有点膨胀起来时，厨师会利索地用手把它拎起来，一下子扔到旁边炉子的火中去，饼被火一烘就膨胀起来，好像悬在火焰上一样，十分奇妙。厨师很快地把它取出，放入盘中，香气四溢。

随着早期印度人移居马来西亚、新加坡等地，他们也将其饮食文化带到了这些国家和地区，同样带去了印度飞饼。在保留基础制作方式的同时，为迎合当地居民的口味而进行了改良。通常，东南亚地区的"印度飞饼"是印度帕拉塔饼的变种，类似食品在马来西亚则称作 roti canai，在新加坡称作 roti prata，在印度尼西亚称作 roti cane，在泰国则称作 ro tee，受到了各国各民族人民的喜爱。

在马来西亚，印度飞饼是一种圆扁形或方形的煎饼，是马来西亚嘛嘛档最受欢迎的一种食物。所谓的嘛嘛指的是当地印度裔的穆斯林居民。传统的嘛嘛档就是嘛嘛在路边经营的饮食档口。印度飞饼可以用来当早餐、午餐、下午茶与晚餐，甚至夜宵。吃印度飞饼时常配着几种达尔酱（素咖喱）或叁巴酱一起吃。

印度飞饼的马来语是 roti canai，这个名字就基本说明了印度飞饼的制作方法。其中 roti 就是面包或饼的意思，也可理解为煎饼；canai 这个词的意思是铺开面团，或者来回压扁，因印度飞饼师傅在制饼过程中把面团压扁，及用"飞"的绝技将面饼拉阔、拉薄。

二 印度飞饼的逸闻趣事

据说在很久以前，印度民间很少有人吃得起白面粉，大多数平民百姓主要以吃粗粮面为主。一天，有位女主人家里来了一位尊贵的客人，可家里只剩下少量的白面粉了，急的主人团团转，不知用什么招待客人好。女主人灵机一动，急中生智，将粗粮面与白面粉掺和在一起，准备烙成饼给客人吃。女主人心想，粗粮面口感差，如果饼烙厚了，其韧性太强，不易咬食，于是她利用面粉的韧性将饼甩得如薄纸一张，再将薄饼折叠煎制。客人吃了烙好的饼后竟赞不绝口。这种煎饼不仅味道特别香美，而且富有韧性，特别是趁热吃时，酥脆可口，越嚼越香。后来，这种饼在印度广为流传，慢慢地由印度人改进做法并带到了东南亚以及世界各地。

三 印度飞饼的分类

印度飞饼按口味可分为原味、甜味、咸味、麻辣味等四大类型；按地域不同可分为马来西亚式印度飞饼（包括水灾飞饼、海啸飞饼、馅料飞饼、其他印度飞饼）、印度尼西亚式印度飞饼（羊肉咖喱作蘸料）、新加坡式印度飞饼、泰式印度飞饼（加入芒果、香蕉等水果，并添加糖、炼乳、果酱或者花生酱等增加甜味）及印度本地飞饼等类型。

四 印度飞饼的制作工艺

（一）主要原辅料介绍

印度飞饼之主料——面粉

面粉是由小麦磨成的粉末，按蛋白质含量的多少，可将面粉分为高筋面粉、中筋面粉、低筋面粉及无筋面粉，印度飞饼常选用高筋面粉。面粉富含蛋白质、碳水化合物、维生素和钙、铁、磷、钾、镁等矿物质，是印度飞饼制作不可少的原料。

印度飞饼之配料——鸡蛋

鸡蛋为印度飞饼提供蛋白质。

印度飞饼之配料——牛奶

牛奶营养丰富、容易消化吸收、物美价廉、食用方便，是“最接近完美的食品”，人称“白色血液”，是理想的天然食品。

印度飞饼之配料——酥油

酥油严格来讲是“起酥油”的一种，是从牛、羊奶中提炼出的脂肪，为印度飞饼提供营养及良好的风味。

（二）酱料的制作

酱料的制作主要通过原辅料筛选，清洗，加水煮制等工序完成。

材料及处理

扁豆 100g，水 800ml，姜黄粉 2g，洋葱 100g（切成方块），胡萝卜 150g，茄子 200g（切丁），番茄 100g（切块），青椒 120g（切块），椰浆 / 牛奶 100ml，盐 3g，罗望子汁 30ml，食用油 15g，茴香子 5g，咖喱叶 1 片，干辣椒 2 个（切段），红葱 3 个（去皮，切碎），蒜头 3 个（去皮，切碎）。

操作要点

1．扁豆洗净，清洗几次，放入锅中，加入 800ml 的水和姜黄粉，煮至沸腾，然后调小火，煮 20min。加入部分洋葱、胡萝卜、茄子和番茄，煮 20min 或直到蔬菜变软。

2．添加椰浆或牛奶，加入少许盐，添加罗望子汁少许；再煮沸后调小火，加少许盐，加入青椒，5min 后熄火，把青椒取出，汤汁备用。

3．用另一个锅，热锅热油，炒茴香子（30s），加入咖喱叶、干辣椒、红葱、蒜头和剩余洋葱，爆香至金黄。然后把炒好的料倒入备用的汤汁中，混合均匀，盖上盖子，加热 5～10min 即可。

（三）基础飞饼的制作

基础飞饼又称“零”飞饼，就是饼内没有加任何馅料的素饼。

基础飞饼制作工艺流程

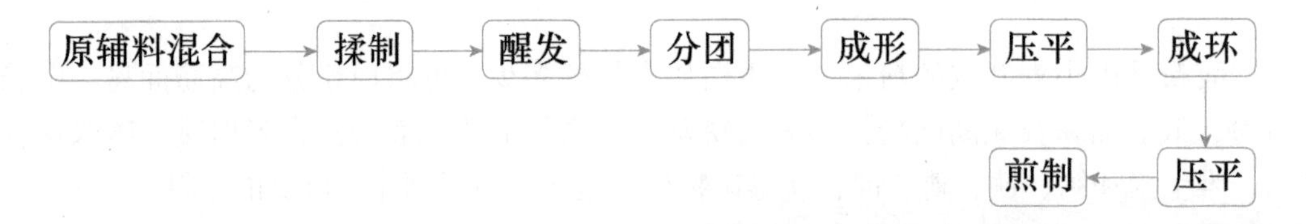

参考配方

高筋面粉 1000g，食盐 6g，温水 / 牛奶 500ml，酥油 200g，鸡蛋 150g，黄油 40g，炼乳 15g，砂糖 10g，色拉油 10g，小苏打 3g。

操作要点

1．混合：把面粉、食盐、鸡蛋、砂糖、酥油与水 / 牛奶混合在一起揉，直到面团柔软光滑。面粉的质量、水的硬度（适当硬度的水中的矿盐能增强面筋）和操作间的温度都会影响用水量。如果加水过量会导致面团粘手，需增加面粉；如果加水太少，会导致揉面困难，需要加水继续揉面。在揉面时加几滴油（植物油）有助于软化面团。

2．揉制：面团揉5～8min，揉得时间越长，面团越好。一定要用力揉，用指关节和手掌底部来拉伸面团，并伸展面筋，以创造正确的纹理。揉好的面团应该是有弹性、不粘手。

3．醒发：放置醒发2～3h或隔夜（最好用布或保鲜膜盖住以免空气进入）。

4．分团：把面团分成16等份小面团。

5．成型：把小面团搓成球形，用擀面杖把小面团压平，拌入酥油10g。

6．成环：卷起面团，扭曲成一个圆环形（将首尾两端粘一起），再揉一遍成圆形，越薄越好，再重复这个动作，卷起面团，扭曲成一个圆环形（将首尾两端粘一起）。

7．压平：用擀面杖把面团压平，压薄一点，以确保空气被赶出。

8．煎制：烧热平底锅，放入少量酥油加热；把压平的面团放在锅上煎，翻到另一面撒一点酥油，煎至两面金黄即可。

（四）鸡蛋飞饼的制作

鸡蛋飞饼制作工艺流程

原辅料混合 → 揉制 → 放置 → 分团 → 抛饼成形 → 加鸡蛋 → 煎制

参考配方

面粉350g，甜炼乳8g，食盐5g，水350ml，黄油或酥油10g，黄油/酥油4g（涂表层），鸡蛋4个。

操作要点

1．制作面团：将水和甜炼乳搅匀，放入盐拌至溶化；放入面粉、蛋黄拌匀成面团。用搅面机将面团快打至起筋，再慢打至柔滑，静置醒发10min，等其柔滑光亮，用手抓一小球（150g左右），抓至光滑，表面涂上酥油放入盆内，醒发3h。

2．抛饼成形：取小面坯，放在特制案板上（可用玻璃板代替，但不可用木制的），抹油，拍成饼，使其慢慢扩大，然后拽住面饼的边缘进行甩动，利用离心力把面饼甩薄，薄到隔着饼几乎可以看到后面的东西，淋浇上少量鸡蛋液，最后将饼一层层叠起。

3．煎制：饼做好后，在平底锅内抹上油，开始煎，待饼的两面煎成金黄色，即可

用铲刀分割成小块，形态可以各异。

五 印度飞饼的风味特色

印度飞饼色泽鲜艳，金黄而透明，口感脆嫩、醇香、味美、可口。风味独特，制作神奇，薄如蝉翼。一般分为两层，外酥里嫩，松软可口；内层绵软白皙，略带甜味，嚼起来层次丰富。一软一脆，口感对比强烈，嚼过之后，齿颊留芳。

飞饼营养丰富，其原料中的牛奶具有一定健胃养脾、生津润肠的作用；酥油有滋润肠胃，和脾温中的作用。

参考文献

贾卫民，王凤成，郑学玲. 2006. 国产小麦粉制作卷饼的适用性研究 [J]. 食品科技，31（09）：54-57.

金婷，谭胜兵，汪成. 2017. 模糊数学法在煎饼感官评定中的应用 [J]. 食品研究与开发，38（03）：25-27.

马涛，王勃. 2012. 煎饼自然发酵糊中酵母菌的分离鉴定 [J]. 食品科技，37（08）：32-35，39.

潘治利，骆洋翔，艾志录，等. 2018. 枣味低油速冻手抓饼的开发 [J]. 粮食与饲料工业，（09）：25-29.

沈嘉禄. 2013. 印度飞饼的游戏精神 [J]. 特别健康，（07）：66.

唐明礼，王勃，刘贺，等. 2015. 煎饼发酵面糊中酵母菌的鉴定及保护剂配方的优化 [J]. 食品工业科技，36（18）：219-223.

王平平，杨芙莲. 2018. 响应面法优化甜荞麦超微粉煎饼的工艺配方 [J]. 粮油食品科技，26（06）：7-13.

肖安红，潘从道，印兆庆. 2001. 添加玉米粉制作发面饼的研究 [J]. 武汉工业学院学报，（01）：4-7.

张元杰. 2018. ELISA 法检测煎饼类食品中黄曲霉毒素含量 [J]. 中国农村卫生，（10）：12-13.

左利民，张璇，崔雷，等. 2017. 专用手抓饼小麦粉的灰分和面筋值对手抓饼品质影响的研究 [J]. 食品与发酵科技，53（03）：61-63.

俄罗斯黑面包

Russian Rye Bread

Russian rye bread, also known as rye bread, is mainly made from rye flour and is one of the staple food in Russia. It is also commonly seen in Western Asia, Northern Europe and Germany. This kind of bread is known as one of the "three major Russian delicacies" along with caviar and vodka.

As early as the ninth century, black bread made from fermented flour had appeared. As an ancient country with a long history, Russia has formed its own unique food culture in the course of historical changes and developments. Just as cereals that have nurtured the Chinese nation for generations, the "black Lieba" is an indispensable food for three meals in Russia from the President to common people.

"Black Lieba" means rye bread in Russian and also refers to grains and cereals. Russian style bread, rooted in the land of Russia, is not only the staple food of the Russians, but also one of their most noble delicacies as it has rich nutrition and unique flavor, and can be consumed conveniently and preserved easily. Therefore, the Russians who had the long tradition of valuing food since ancient times have formed their own unique "bread culture" in their long-term labor and life practice. Among all kinds of Russian bread, rye bread is typical and very popular among Russians.

What rye bread to Russia is what dumplings to China, red wines to France, and sausages to Germany. It has become a symbol of Russia. Bread is a traditional staple food for Russians. You have a wide variety of bread products to choose from in Russian shops.

Rye bread is a special food for Russian army. During the World War Ⅱ, it was the lifeline for the Soviet army. Rye bread saved at least 400 000 lives in Russia and sustained the combat capacity of nearly 10 million people.

Leningrad siege was the longest and most destructive bloody war which caused the second largest death toll in modern history. In the days when Leningrad was besieged by the German army, the only food available to the civilians in Leningrad was 125 grams of rye bread per person according to the maximum supply capacity. With the 125 grams of rye bread, the Leningrad people lived through a difficult period of 872 days. The 125 grams of rye bread was their lifeline. Even today, if you talk to the Russians about "125 grams of rye bread", it will remind them of the Leningrad Defence War. 125 grams of rye bread is almost the synonym of Leningrad Defence War.

In a corner of the panoramic display hall concerning the Leningrad Defence War in the World War Ⅱ History Memorial Hall in Russia, you can see a picture of an old woman getting 125 grams of rye bread in a tall counter window, which is the same tense and solemn as withdrawing huge sums from the bank.

"Bread and salt" are the most precious food for Russians and represent Russian's cultural connotation. They are highly respected and admired in the eyes of Russians and have a very important symbolic meaning: bread stands for harvest and wealth while salt can protect people from evil. So treating guests with bread and salt has been an ancient custom in Russia. At the beginning or end of each meal, a slice of bread with a little salt is eaten as a sign of good luck. Serving "bread and salt" to guests will increase the trust and friendship between the guest and the host. If someone refuses to enjoy it, it means disregard and insult to the host. In the folk, when meeting others during the dinner, people always say "bread and salt" to each other with enthusiasm. It is said that this phrase can help them drive away evils and pray for peace. "Bread and salt" are not only a showcase of the host's generosity and hospitality, but also can remove people's suspicion and hostility at a critical time.

The rye bread is black in color, rich in dietary fiber, and easy to be digested. It tastes slightly acidic and salty and has a distinctive flavor. The surface is often wrapped with buckwheat husks, wheat kernels, pumpkin seeds, sesame seeds, other plant seeds, etc. so as to increase the flavor and taste. Rye bread is chewier than wheat or barley bread and has a stronger taste.

俄罗斯黑面包是用黑麦和在磨粉过程中被碾下来的皮层、胚芽、糊粉层及少量的胚乳等麸皮，经过酸面团发酵制作而成，为俄罗斯的主食之一，在西亚、北欧以及德国都较为常见，与鱼子酱、伏特加一起被誉为俄罗斯美食“三剑客”。

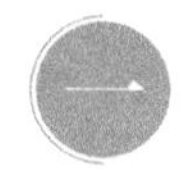 黑面包的起源和发展

早在9世纪的时候就已经出现了用发酵面粉烘烤而成的黑面包。俄罗斯作为一个历史悠久的文明古国，在长达千余年的历史变迁和发展过程中，形成了自己民族特有的饮食文化。正像五谷杂粮世世代代养育着中华民族一样，在俄罗斯上至国家总统，下至平民百姓，“黑列巴”一日三餐不可缺少。

在俄语中，“黑列巴”是黑面包的意思，也泛指粮食和谷物。根植于俄罗斯大地的俄式面包，因其营养丰富、风味独特、食用便利、容易保存等特点，既是俄罗斯人的主食，也是他们最尊贵的一种美食。自古以来就珍爱粮食的俄罗斯人在长期的劳动生活中形成了自己民族所特有的“面包文化”，种类繁多的俄式面包中尤以黑面包备受俄罗斯人的青睐。

 黑面包的逸闻趣事

（一）125克黑面包

黑面包是俄罗斯军队的特色食品。第二次世界大战（简称“二战”）时期，黑面包更是苏联军队的保命干粮。黑面包至少挽救了俄罗斯40万人的生命，维持了将近1000万人的战斗力。

列宁格勒围城战是近代历史上主要城市遭受封锁时间最长、破坏性最大和死亡人数第二多的血腥战役。在德军重兵围困列宁格勒的日子里，按最大的供应能力，列宁格勒的平民每天所能得到的唯一食物——125克黑面包。靠着这125克黑面包，英勇的列宁格勒人民度过了艰难的872天，它是列宁格勒人支撑生命的底线。即使在今天，如果你对俄罗斯人说起“125克黑面包”，他们的反应一定是列宁格勒保卫战。125克黑面包几乎可以和列宁格勒保卫战画上等号。

在俄罗斯二战历史纪念馆关于列宁格勒保卫战的全景画厅的一个角落里，你可以看到一位老妇人在一个高高的柜台窗口里领取125克黑面包的画面，那个场景就像是在银行领取巨款一样紧张和凝重。

（二）黑面包的文化——面包和盐的待客礼俗

“面包和盐”是俄罗斯人最珍贵的食物，也是最具民族内涵的食物，在俄罗斯人的心目中备受尊敬与崇尚，并具有非常重要的象征意义：面包代表着丰收和富裕，盐则有避邪之说。因此，在俄罗斯形成了用面包和盐招待宾客的古老习俗。在每餐开始或者结束的时候，大家都会吃上一片加少许食盐的面包，以示吉祥。向客人奉上“面包和盐”会增加宾主之间的信任与友谊。如果对方拒绝享用，则意味着对主人的漠视与侮辱。在民间，吃饭时碰见旁人，人们总是热情地冲着对方说“面包和盐”，据说这个词组能帮助他们驱走恶神，祈福平安。

至今俄罗斯人仍延续着这一传统，将面包和盐放在一起，表示对客人最热烈的欢迎，也表达对客人的亲切、友好和尊重之情。现在，俄罗斯国家领导人到外地视察工作时，当地政府也经常以这种仪式迎接。另外，这一传统仪式也常用于隆重的外交场合，以表示对来宾的热烈欢迎及两国人民深厚的友谊。

黑面包制作工艺

（一）原辅料介绍

黑面包种类丰富且繁杂，不易分类，但其主要成分相差无几，一般由黑麦粉、天然酵母菌种、麦麸、盐作为主料，可根据不同品种或口味需求调节黑麦粉、酵母菌及水的比例，再辅之以糖、麦仁、香草籽、芝麻、荞麦皮调味。

黑面包之主料——黑麦粉

黑麦富有营养，含淀粉、脂肪和蛋白质、维生素B和磷、钾等，对人体非常有益。黑麦粉是制作黑面包的主要原料，其赋予了黑面包独特的风味和颜色。

黑面包之辅料——天然酵母菌种

酵母菌可利用其生命活动过程中所产生的CO_2和其他成分，使面团发酵从而膨松并富有弹性，并赋予成品特殊的色、香、味及多孔性结构。一般使用天然酵种进行黑面包的发酵，但耗时较长，且酵种的选择与发酵时间对黑面包的酸度与松软度有很大影响。

黑面包之辅料——麦麸

麦麸含有非常丰富的膳食纤维，可提高食物中的纤维成分。在黑面包的制作中适当添加少许麦麸对人体是非常有益的，且对营造黑面包独特的口感起到一定的作用。

黑面包之调味料——糖、麦仁、香草籽、芝麻、荞麦皮

黑面包口感偏酸，加入糖可大大增加其甜味，使其口感香甜，且糖粒经高温烘烤熔化后，可造成面包表面的麻点及内部大孔洞，一定程度上增加了黑面包的松软度。荞麦皮、麦仁、南瓜子、芝麻等不仅可以增加面包的香味和口感，使黑面包更香脆可口，且其本身的营养成分还可增加黑面包的附加营养价值。

（二）黑面包的制作

黑面包制作工艺流程

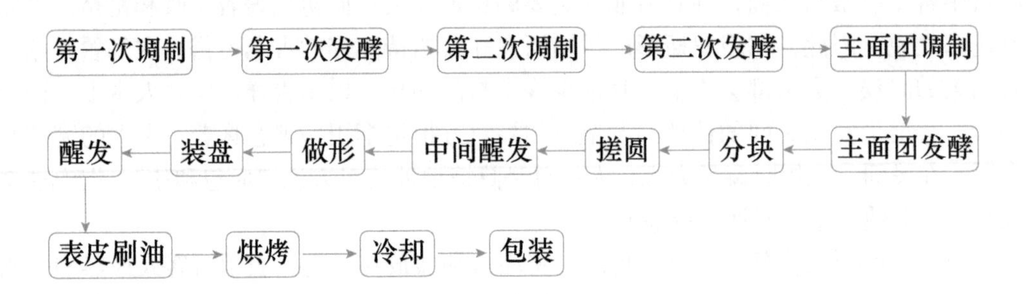

参考配方

粗磨黑麦粉500g、小麦粉500g、糖120g、麦麸30g、鲜酵母15g、水450ml、酪浆（buttermilk）或乳清（whey）250ml、食盐7.5g及调味料（香草籽、芝麻粒、麦仁等）适量。

操作要点

1．种子面团的第一次调制及发酵：种子面团调制和面团发酵是密切相关的两个工序。第一次调制面团用粉量为全部黑麦粉的15%，即黑麦粉75g以及酵母2g、水45ml，将原辅料放入和面机中，慢速搅拌3min，中速搅拌5min成面团，在25～28℃下发酵20～24h。

2．种子面团第二次调制及发酵：将第一次发酵好的种子面团掺入30%～40%的黑麦粉（150～200g）搅拌均匀，调成有弹性的面团，于温度29～32℃、相对湿度75%的条件下发酵18～20h。随时注意发酵程度，待发酵至内部布满小孔，发酵的气体冲破表面时即可进行下一次发酵。

3．主面团调制及发酵：将第二次发酵的面团与剩余的黑麦粉、小麦粉、酵母、糖加入和面机中，同时加入食盐（以盐水形式加入）、麦麸及调味料（香草籽、芝麻粒、麦仁等），慢速搅拌5min，中速搅拌10～12min，使之形成面团，最后加入酪浆或乳清，继续搅拌5min，于25～37℃的环境进行主面团发酵，发酵时长为2～4h。

主面团调制主要影响面团的弹性与伸展性，对黑面包的口感有很大影响。调制面团时需使各种原辅料均匀分布，注意搅拌力度，使面团具有良好的弹性和伸展性，改善面团的加工性能；使面团变得柔软且处于易伸展的状态，即面团膜薄层化。

4．分块与搓圆：将面团按所需大小进行分块。将面团分块时，面团发酵仍然在进行中，因此要求面团的分割时间越短越好，最理想是在15～25min以内完成。将分割后的不规则小块面团搓成圆球状，搓圆完成后一定要注意收口向下放置，避免面团在醒发或烘烤时收口向上翻起形成表面的皱褶或裂口。

5．中间醒发：醒发温度40℃左右，湿度85%～90%，时间50min左右。湿度过低，面包坯表面容易结皮干裂；温度过高，面包坯表面容易结露水，产生斑点，甚至塌架；时间不足，烤出的面包体积小；时间过长，面包酸度增大。

6．做形、装盘、醒发：将经过中间醒发后的面团，按照面包的品种要求，制做成不同形状，这一过程称为做形。做形的要求是在不损坏面团的情况下，驱赶面团中的二氧化碳气体，充分混入新鲜空气，使面团中的酵母发酵得更好，并做成各种形状的面团，将面团装入模具进入最后醒发。最后醒发温度38～40℃，相对湿度80%～85%，时间45～55min。一般醒发至面团膨胀到原体积的2～3倍为止。

7．表皮刷油：表皮刷黄油是为了防止面包糊底与粘盘，且可使面包表面富有光泽，增加面包的色香味。除面包表皮之外，烤盘也要刷油。

8．烘烤：烘烤是保证面包质量的关键工序。黑面包坯在烘烤过程中，受炉内高温作用，组织膨松，富有弹性，表面呈褐色，由生变熟，产生可口的香甜气味。

黑面包烘烤一般分为三个阶段：膨胀阶段，面火120℃，底火200～220℃，10min；成熟阶段，面火270℃，底火200～220℃，10min；上色阶段，面火180～200℃，底火180～200℃，40min。

9．冷却与包装：冷却是为了防止面包变形和防止面包霉变。冷却温度22～26℃，相对湿度75%。家庭制作一般采用自然冷却法。工厂制作黑面包，为了节省时间成本多使用强制冷却法。包装的作用是保持面包清洁，防止面包变硬。待面包冷却后，密封保存。

四 黑面包的风味特色

黑面包颜色偏黑，富含膳食纤维，口感微酸带咸，香气独特，质地厚重，颇有咀嚼感，别具风味。黑面包表面常裹荞麦皮、麦仁、南瓜子、芝麻等，可增加香味和口感。

黑面包的主要原料是黑麦粉，它含有丰富的维生素（维生素B_1、维生素B_2、维生素C、维生素A和维生素E），营养价值很高，能够完善日常饮食中摄入不足的维生素种类，进而保持人体营养的均衡。黑面包中含丰富的阿魏酸和膳食纤维，具有一定的抗氧化作用。另外，其中含有的β-葡聚糖对预防高血压、提高肌体免疫力、降低心血管发病率等也有一定帮助。

参考文献

李新华，董海洲．2009．粮油加工学［M］．北京：中国农业大学出版社：92-97．

全国工商联烘焙业公会. 2009. 中华烘焙食品大辞典（产品及工艺分册）[M]. 北京：中国轻工业出版社.
申瑞玲，祝红蕾，李红. 黑麦的营养保健功能及其在食品中的应用 [J]. 河南工业大学学报（自然科学版），29（05）：79-82，92.
徐明高. 1992. 甘肃省食物营养成分表 [M]. 兰州：甘肃民族出版社.
张怀珠. 2014. 食品工艺 [M]. 北京：中国农业出版社：24-37.
张慧，温纪平，郭林桦，等. 2014. 黑麦营养特性及其在食品中的研究 [J]. 食品研究与开发，（03）：97-99.

皮塔饼

Pita

Pita, also known as Arabic bread or pocket bread, is a round and pocket-shaped flour-made food popular in Greece, Turkey, the Balkans, the eastern Mediterranean and the Arabian Peninsula.

Pita is baked at a comparatively high temperature (about 800 to 900 °F) where the thick dough swells significantly by the force of the steam and turns flat after it cools with an internal pocket left. Pita is cut lengthwise from the middle. Some ingredients and dishes, such as smoked chicken, ham, fried eggs, bacon, fresh vegetables, pickles, and so on, can be stuffed into the pocket-shaped bread.

Made from wheat flour, the soft and slightly puffed Pita takes its roots from the Near East, presumably Mesopotamia, around 2500 BC. It slightly resembles other puffed bread, like South Asian flatbread, pizza crust and Central Asian naan.

From what archaeologists have found out, Pita stemmed from groups of people in the west of the Mediterranean. It was not entirely clear that whether it was the Bedouins or the Amorites who were the pioneers. Pita was considered as an invention in the framing society. Before long, as the Bedouin groups had increased their trade in goods and services, Pita traveled across the Arabian and Sahara deserts. Later, its popularity surged worldwide since it spread eastward along with the Greek culture into the Arabian Peninsula, India, and Afghanistan.

Following the modern Greek word "πίτα", the word Pita derived from the Byzantine Greek "bread, cake, pie, pitta" and appeared in English for the first time. Initially, the pita was a blend of batter that was left to ferment and then yeast was applied to speed up the fermentation. In the Middle East, it is still regularly made in a patio stove.

In the 1970s, Pita became popular in the western world for its interesting feature of being a pocket. People put various ingredients into the pocket and created Pita with stuffings, which were affectionately called "Pita pie". Up to now, in addition to sandwiches, burgers and rice balls, Pita is very favorable in the breakfast market and the fast-food industries for its pocket shape, which can be sliced and served with a sauce or dishes such as chicken, steak and lamb.

When baked, the thick dough of pita swells significantly with an internal hallow pocket (formed by steam) . It is called "pocket bread" for its resemblance to a pocket. The Pita has light golden yellow color, soft texture, and chewy taste.

皮塔饼（Pita）又称阿拉伯薄面包或口袋饼，是一种圆形口袋状面食，广泛流行于希腊、土耳其等地区和阿拉伯半岛。

皮塔饼的起源及发展

皮塔饼是在较高的温度下烘烤制成的，其“口袋”由蒸汽膨胀形成，面饼冷却后变得平坦，中间留下一个口袋。从中间纵切便成为一个口袋形状的面包，可塞入任何食材和菜肴，如熏鸡肉、火腿、煎蛋、培根、新鲜蔬菜、酸黄瓜等。

皮塔饼是由小麦面粉制成的软而略微膨松的面饼，起源于公元前2500年左右。皮塔饼与其他膨化的面包有些相似，如比萨饼皮、南亚的扁面包、伊朗大饼和中亚的馕。

据考古学家解码，皮塔饼的发明者是源于地中海的贝都因人或亚摩利人。随着贝都因人商贸和服务业的发展，皮塔饼穿越阿拉伯和撒哈拉沙漠。后来随希腊文化向东传播，这种美食传到了阿拉伯半岛、印度和阿富汗，逐渐在世界范围声名鹊起。

皮塔饼首次在英语中出现，是沿袭现代希腊语“πίτα”，源自拜占庭希腊语“面包、蛋糕、馅饼”。最初，皮塔饼仅仅是将面糊自然发酵，之后发展到采用酿造酵母进行发酵。在中东，皮塔饼仍然是在天井炉中烘烤制成的。

20世纪70年代，皮塔饼开始在西方世界流行，就是因为“口袋”这种有趣的特点，人们把多种食材填入口袋，做成了带馅料的皮塔饼，并亲切地管它叫“皮塔派”。现今的早餐食品除了三明治、汉堡、饭团之外，皮塔饼因形状酷似口袋，可以切成块，蘸酱或者搭配鸡肉、牛排、羊肉等菜肴，在早餐市场和速食餐饮行业也很受欢迎。

皮塔饼的填充方法

皮塔饼最大的特点莫过于烤的时候面团会鼓起来，形成一个中空的面饼，看着跟一个口袋似的。所以，皮塔饼也被称为“口袋面包”。其口袋可用于填充其他材料，丰富皮塔饼的滋味。

沙拉三明治（炸鹰嘴豆）皮塔饼

皮塔饼中的经典，沙拉三明治（炸鹰嘴豆）皮塔饼是以炸鹰嘴豆为馅料，通常搭配西红柿、切丁黄瓜和切片的洋葱，整个皮塔饼配上芝麻酱汁，适合午餐或晚餐。

萨比奇

萨比奇几乎与沙拉三明治皮塔饼一样受欢迎。萨比奇属于中东三明治，填充材料有炸茄子、煮鸡蛋、鹰嘴豆泥、芝麻酱、以色列沙拉（西红柿丁、黄瓜丁和切片洋葱）以及一种叫作“amba”的甜辣芒果酱。其将很多不同的口味，都完美地结合在一起。

沙瓦玛三明治

沙瓦玛三明治是中东的旅行餐，用鸡肉、牛肉、羊肉甚至素食制成。经典沙瓦玛三明治，是将薄切鸡肉包裹在皮塔饼中，并包括蔬菜、调味料和酱汁。

三 皮塔饼的制作工艺

（一）原辅料介绍

皮塔饼的制作原料主要有面粉、干酵母、橄榄油、食盐、绵白糖等。

皮塔饼之主料——面粉

制作皮塔饼的面粉有高筋面粉和全麦面粉，其中高筋面粉可使面饼具有较好的韧性、弹性，使体积膨大；全麦面粉含有丰富粗纤维、B族维生素、维生素E及锌、钾等矿物质，可赋予皮塔饼微褐的颜色，粗糙的质地以及丰富的麦香。

皮塔饼之发酵剂——干酵母

酵母属于真菌，富含蛋白质、维生素、矿物质和膳食纤维等营养成分。在皮塔饼面饼发酵过程中，通过消化面粉中的糖类物质产生气体和酒精，使面团膨松、具有香味。同时，酵母的各种营养物质留在发酵的面中，增加其营养，对人体健康有利。干酵母的用量一定要适中，过多，烤出来的面包会带酸味；过少，面包成品会不膨松，而且难以消化。

皮塔饼之辅料——橄榄油

橄榄油在地中海沿岸国家有几千年的历史，在西方被誉为“液体黄金”“植物油皇后”“地中海甘露”，原因就在于其极佳的营养价值。因其由新鲜的油橄榄果实直接冷榨而成的，不经加热和化学处理，保留了天然营养成分。皮塔饼制作过程中，以橄榄油刷烤盘，避免面饼与烤盘间粘连。

皮塔饼之调味——绵白糖

绵白糖在皮塔饼制作过程中，提供酵母生活的营养物质，同时调整面团中的面筋胀润度，改进成品组织状态，增加色泽和风味，使产品具有甜味并提高营养价值。

（二）皮塔饼的制作

参考配方

高筋面粉300g，全麦面粉200g，黄油50g，鸡蛋75g，干酵母5g，绵白糖125g，食盐5g，水180g。

皮塔饼制作工艺流程

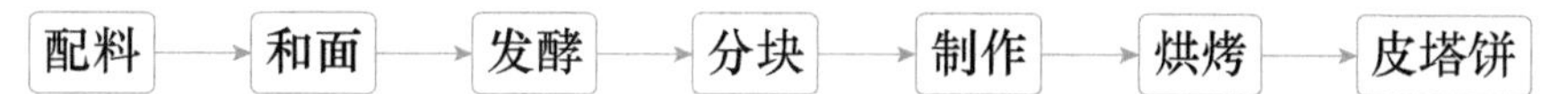

操作要点

1．配料：将面粉（高筋面粉：全麦面粉＝3：2）、绵白糖、食盐、干酵母等粉状称好，放置于搅拌桶中搅拌均匀待用。

2．和面：水加入桶中，先用慢速搅拌2min，再中速12min、慢速2min、中速4min，搅拌至面团中面筋完全扩展为止，以容纳更多酵母发酵所产生的二氧化碳气体，在烘焙阶段使产品体积膨胀更大。

3．发酵：面团发酵温度控制在38℃、湿度80%～85%，发酵40min。温度不宜太低，否则会影响面团的柔软度，进而影响烤焙后体积的膨胀。

4．分块：每个面团分割为60～70g，滚圆，醒发15min后，按压至厚度为2.2～2.5mm。

5．烤制：将烤箱上下火230℃预热，烤盘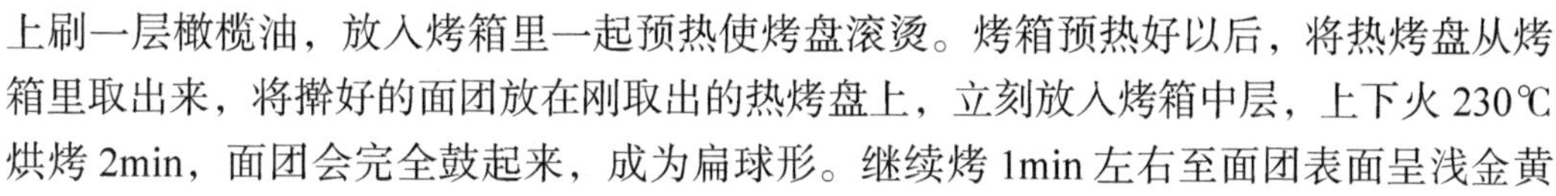上刷一层橄榄油，放入烤箱里一起预热使烤盘滚烫。烤箱预热好以后，将热烤盘从烤箱里取出来，将擀好的面团放在刚取出的热烤盘上，立刻放入烤箱中层，上下火230℃烘烤2min，面团会完全鼓起来，成为扁球形。继续烤1min左右至面团表面呈浅金黄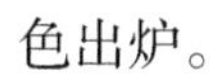色出炉。

经冷却后，须立即包装以防止水分散失影响产品品质，也可冷冻贮存以维持品质。

四 皮塔饼的风味特色

皮塔饼烤的时候面团会鼓起来，形成一个中空的面饼。其色泽呈浅金黄色，质地松软，口感较韧。

皮塔饼富含碳水化合物，能为机体提供能量，具有增加饱腹感的作用；辅以完美搭档鹰嘴豆，使其还有一定的美肤、丰乳、降血脂、减肥等作用。

参考文献

陈中爱，董楠，陈朝军，等. 2018. 含不同粗粮粉面包的营养、质构特性、风味化合物［J］. 食品工业科技，39（04）：21-27，32.

李勇，苏世彦. 1997. 冷冻面团焙烤面包的生产技术［J］. 食品科技，（02）：15-16.

杨鹏，袁梦，于博，等. 2019. 传统烤饼复合保鲜剂的研制［J］. 中国食品添加剂，30（04）：119-126.

Jones D. 2011. Cambridge English Pronouncing Dictionary. 18th ed. [M]. Cambridge: Cambridge University Press.

Pasqualone A. 2018. Traditional flat breads spread from the Fertile Crescent: productive process and history of baking systems [J]. Journal of Ethnic Foods, 5 (1): 10-19.

匈牙利姜饼

Hungarian Gingersnap

Gingersnap is a type of cookies specifically for Christmas. It is made of honey, brown sugar, almond, preserved fruit and spice. There are two English names for this food. One is gingersnap, which means a ginger plus thin and crispy bread. The other is Lebkuchen from German. Gingersnap can be made into different forms such as chocolate gingersnap and gingersnap cottage.

In Europe, the gingersnap has existed for many years. Long before Christmas became a commercial holiday for the entire public to celebrate, gingersnap had been a part of winter rituals in Ancient Europe. Before the mid-century, people had the tradition of celebrating the winter solstice for thousands of years and as time moved on, it had been replaced by Christmas in most European reigions while the old feast tradition remained.

Gingersnap is a classic Christmas cookie and the flavor remains the same as it was in mid-century. However, there was no such commonly seen gingersnap man at that time. The first person who made gingersnap is Elizabeth I, the Queen of England, and she turned them into the shape of the minister she loved most.

Gingersnap is an ancient food. It could be traced back to a cuisine book named *Cooking and Eating in the Roman Empire* (a book that introduces the cooking methods and table manners in Rome Empire from 27 BC to AD 1453), which was published in Roman Empire period. Ginger, an important ingredient of the gingersnap, was extremely popular and widely planted at that time. In the mids-century, since gingersnap could be preserved for a long period, many young noble females sent gingersnap decorated with gold as a gift to the knight who was going to participate in an equestrian event. As a result, gingersnap became a symbolic item and many bazaars themed by gingersnap appeared. Some single females from the village ate husband-shape gingersnap to deliver a wish for an appropriate spouse. In 1796, the first publicly published cuisine book in USA–*American Cuisine*, mentioned four different methods to cook gingersnap.

There was a dessert shop in Magler square in Hungary and it was owned by an old lady. Gingersnap was the only item that had been sold in the shop. She only cooked gingersnap in a boy's shape rather than a girl's shape because she sold gingersnap as a token of her yearning. When she was young, the World War broke out. She saved an officer and they felt in love. But unfortunately, it turned out that the officer was a spy from the enemy. She kept the secret as she loved him very much. At the Christmas of that year, the old lady cooked a pair of gingersnap, a boy and a girl. When she decided to give these gingersnaps to the officer, she found that someone snitched the officer and he had to run. Before he left, she gave the girl-shaped gingersnap to the officer. As decades passed by, she had never met the officer. So, the old lady had been cooking human-shaped gingersnap in memory of her lover.

"Ginger" was an expensive import spice. Therefore, it only could be used on an important holiday such as Christmas and Easter holiday. The chefs put ginger into cake and cookie to enhance the flavor and may increase the function of against coldness. As time went on, gingersnap became a dessert that linked to Christmas. Later, there was a type of "gingersnap Bazaar" in Europe and different shapes of gingersnap were offered in different seasons. Among them, the most famous one was from Nuremberg, Germany. It was praised as the "Capital of gingersnap". On 6th December, the San Nicolas Day, godfather and godmother in Northern France and Germany sent different shapes of gingersnap, such as heart-shape and human-shape to children or secretly put them in the socks. Gradually gingersnap became part of the fairy tale and was cooked into different shapes.

Traditional gingersnap uses honey and pepper powder as the ingredients. It tastes both spicy and sweet with irritating flavor. Nowadays, gingersnap has been improved with a cover of icing to enhance its taste together with a better appearance.

姜饼是圣诞节时吃的小酥饼，通常用蜂蜜、面粉、糖及香辛料制成。姜饼的英文有两种，一种是Gingersnap，可以理解为姜加薄而脆的饼；而另一种说法来自德语，叫Lebkuchen。姜饼有很多种花样，如巧克力姜饼、姜饼小屋等。

姜饼的起源及发展

在欧洲，圣诞姜饼已经流传了许多年。早在圣诞节变成如今这种全民欢乐的商业节日之前的很长一段时间，姜饼就已经是古代欧洲冬至仪式的一部分了。在中世纪前，人们庆祝冬至的传统已经有几千年的历史。到了中世纪，欧洲的大部分地区已经用圣诞节取代了冬至，但旧的筵席传统却保留了下来。

姜饼是一款经典的圣诞饼干，现在的味道和中世纪时的基本一样，不过当时没有现在常见的姜饼小人。第一个做姜饼小人的，正是英国女王伊丽莎白一世，她把饼干做成了她最喜爱的大臣的形状。

关于姜饼最早的记录可见于古罗马帝国时代一本叫《罗马帝国的烹饪和用餐》的烹饪书。姜饼的主要原料——姜，在当时非常受欢迎而且种植广泛。在中世纪时，因姜饼能保存很久，许多年轻的贵族女性会把用真金装饰过的姜饼，送给即将参加马术的骑士。因此姜饼成了一种很具代表的象征，也产生了许多以姜饼为主题的市集。有些村庄的未婚女子在市集上吃“husband”形状的姜饼，期望找到适合自己的配偶。1796年，第一本在美国公开印刷发行的食谱书——《美国的烹饪》，书中提到了四种姜饼的做法。

在过去，姜只用在圣诞节、复活节这样的重要节日，把姜加入蛋糕、饼干中可以增加风味，并有驱寒的功用。久而久之，姜饼就成了与圣诞节关联的点心。后来，在欧洲出现了一种“姜饼市集”，市集在不同的季节提供不同形状的姜饼。其中德国的纽伦堡更享有“姜饼之都”的美誉。在每年的12月6日圣尼古拉斯节时，在法国北部和德国，教父、教母都会在这一天送各种形状（如心形、人形）的姜饼给孩子们，或偷偷地放入孩子们期待礼物的袜子内。

姜饼的逸闻趣事

在匈牙利麦格勒广场有一间甜品屋，甜品屋的老板是个老奶奶，她只卖一种食品——姜饼。只做男孩形状的，不做女孩形状的。因为，她卖的是寄托心思。原来在老奶奶年轻时，遇上了世界大战。她救了一个军官，后来他们相爱了。但是，她发现，对方原来是敌方间谍。不过她爱他，并没有说出去。在那年圣诞节，老奶奶亲手做了一对姜饼人，一个男孩和一个女孩。可当她拿给军人时，发现心上人被告密了，必须要逃亡。临走前，老奶奶把女孩姜饼交给了他。几十年过去了，老奶奶再也没见到那位军人。老奶奶就用做的姜

饼人来纪念她的爱人。

 姜饼的制作工艺

（一）原辅料介绍

姜饼主要原辅料有低筋面粉、酥油、绵白糖、蛋清、小苏打粉、泡打粉、姜粉、肉桂粉、丁香粉和蜂蜜等。

姜饼之主料——低筋面粉

低筋面粉是指水分小于13.8%，粗蛋白质9.5%以下的面粉，通常用来做蛋糕、饼干、小西饼点心、酥皮类点心等。其赋予姜饼良好质地。

姜饼之辅料

1．酥油：酥油具有一定的可塑性和稠度，用作糕点的配料、表面喷涂或脱模等。主要作用是酥化或软化姜饼，并改善口感。

2．泡打粉：泡打粉是一种复合膨松剂，用于姜饼可增大体积，产生膨松感。

3．肉桂粉：肉桂粉由肉桂或大叶清化桂的干皮和枝皮制成的粉末，用于姜饼产生芳香味。

4．丁香粉：气味强烈、芳香浓郁、味辛辣麻，可用在各类调味料中，也可制成香精或入药。在姜饼中可矫味增香。

5．蜂蜜：蜂蜜常用于增加口感，且易被人体吸收，可增加姜饼的滋味。

6．蛋清：蛋清的固形物中大约90%是蛋白质，在面粉中加入蛋清不仅能增加面筋劲力，还能提高姜饼的营养价值。

7．姜粉：加入姜粉可以增加姜饼风味，并有驱寒的功用。

（二）姜饼的制作

姜饼制作工艺流程

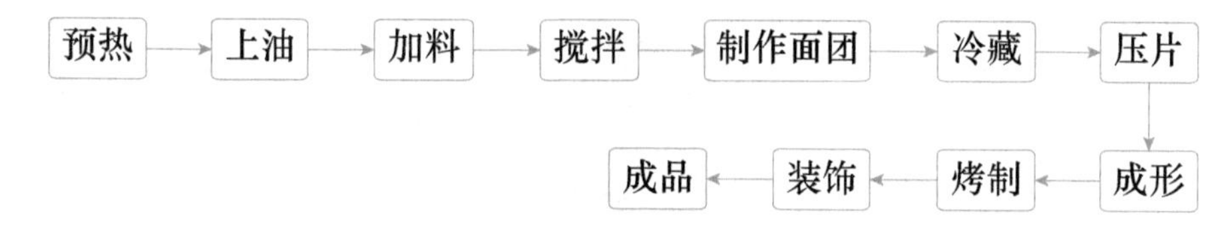

参考配方

低筋面粉700g、酥油1200g、绵白糖90g、蛋清40g、小苏打粉5g、泡打粉6g、姜粉10g、肉桂粉4g、丁香粉4g和蜂蜜100g等。

操作要点

1．预热、上油：先预热烤箱至190℃，烤盘涂上一层油备用。

2．加料、搅拌：酥油放入电动搅拌器中，中高速搅拌30s，加入绵白糖、泡打粉、

姜粉、小苏打粉、肉桂粉和丁香粉，全部搅拌均匀。

3. 制作面团、冷藏：加入蜂蜜、鸡蛋等，混合均匀，最后再加入面粉混合成面团。将搅拌好的面团放入冰箱冷藏3h以上，或是冰至面团不粘手可以很容易擀开的程度。

4. 压片、成形：在案板上撒薄薄一层干面粉，把保持低温冰好的面团，快速擀成片，并用饼干模型，压成各式形状。

5. 烤制、装饰：把面团放入烤箱烤5～6min，烤至饼干边缘呈金黄色即可。从烤箱取出烤盘，略放冷却，最后再用巧克力或糖霜装饰冷却好的饼干。

（三）姜饼屋的制作

姜饼屋制作工艺流程

绘制草图 → 制作饼干 → 拼接、粘合 → 装饰成品

配料及工具

制作姜饼屋的原辅料与姜饼制作工艺所需原辅料基本相同，除此以外，还需要糖霜、色素及饼屋装饰制作工具。

1. 糖霜：主要成分是白糖、玉米淀粉、葡萄糖、麦芽糊精、卵磷脂、色素等，因颜色鲜艳，可用于提高姜饼屋装饰效果。

2. 色素：食用色素可使姜饼在一定程度上改变原有的颜色，用于姜饼屋色彩的装饰。

3. 制作工具：美工刀和尺子（用于姜饼屋草图的绘制）、硬卡纸（用于雕刻姜饼屋的墙体和顶棚）、裱花袋（用于盛装制作好的糖霜，起到给姜饼屋上色的作用）。

操作要点

1. 草图绘制：在硬纸板上绘制姜饼屋草图并用美工刀进行剪裁。对于不对称的房屋结构，在切割饼干坯时要注意正反面。

2. 饼干制作：黄油融化，加白糖搅拌均匀，再筛入所有粉类原辅料，揉匀。分小份入袋中按

扁，冷藏30～60min至稍硬；取出一份，放在油纸上，先擀一个大的底座，约3mm厚，大小根据需要而定。用滚针插孔，防止烤时鼓起。直接把油纸移入烤盘，170℃烘烤10min。取出烤盘放烤网上晾凉，用裁剪好的硬卡片在油纸上分别刻出墙体顶棚，把油纸移入烤盘入烤箱烘烤。饼干烤好后放置一边放凉；取块糖，砸碎后放入烤好的饼干窗户里，入烤箱烤至融化，凉后取下即成窗户。

3．拼接、粘合：把饼干晾置一夜，备好料和工具，第二天开始搭房子。蛋清里分三次加入糖粉，打至硬性。提起打蛋器时不跌落，盆内痕迹不会消失即成糖霜。然后把裱花袋套在高杯子上，把糖霜盛入即可；同时取三个小碗，分别盛入两勺糖霜，分别加入几滴色素，即成三色糖霜。房子制作时，先在底座上合适的位置挤糖霜，粘上墙体和门，放置一夜，待墙体完全干后，粘上顶。

4．装饰成品：用彩色糖霜进行装饰，放置一夜，糖霜基本能干透。

 姜饼的风味特色

传统的姜饼，以生姜、蜂蜜为材料，具有甜辣味，口感刺激；现在的姜饼经过改良，外面撒上一层糖霜，不但口感丰富，外形也美观。

姜饼采用生姜为原料，其性温、味甘、入脾，有助于益气补血、健脾暖胃、缓中止痛、活血化瘀。

参 考 文 献

杜亚军，杨春，张江宁，等．2017．全谷燕麦香酥饼的工艺研究［J］．食品科技，42（02）：181-186．
李新华，董海洲．2009．粮油加工学［M］．北京：中国农业大学出版社：98-100．
童金华，黄培姗，蔡若萍，等．2017．特色风味酥饼加工工艺研究［J］．福建轻纺，（01）：33-38．
吴姝宓．2018．乳清奶酪酥饼的加工工艺及其品质功能特性的研究［D］．呼和浩特：内蒙古农业大学．
徐明高．1992．甘表肃省食物营养成分表［M］．兰州：甘肃民族出版社．
张怀珠．2014．食品工艺［M］．北京：中国农业出版社：44-47．

派

Pie

Pie is a kind of dessert consisting of filling and pastry. Due to its simple making method and rich nutrition, pie is popular among the people in Europe and the United States. Pie can be made to different varieties with different fillings and pastries. Apple pie, the most classic type of pie, has become a common dessert in American's life.

Pie originated in Europe with the filling of apples. Five hundred years ago, when Christianity faced the challenge from the Islamic world, people lacked food and thus planted the taboo fruit "Apple"of Christianity in large areas. When the Sudanese army retreated, a large amount of harvested apples became a burden. It was the best choice to eat roasted apples as the main course with knives and forks because eating raw apple was considered as a taboo. The roasted apple was soft and delicious and the taste was sweeter after roasting, thus becoming the best substitute for starchy food. Later the French opened the bottom of apple, removed the core, and filled the inside with butter and sugar. After being baked, the butter can add the taste of meat-like thickness after melting while sugar can increase sweetness. The Spaniard was even more creative. In addition to butter and sugar, they also added red wine and nuts inside apples, and even poured caramel and chocolate on the surface of baked apples. People in the colder north tended to add starch in roasted apples as the staple food, which became the earliest apple pie.

As baking of fruits became popular in southern Europe, apple pie spread to northern Europe. Sweden was the first Nordic country to fill apples in a pie. The Gothic pie had no bottom, and its wrapper was not an even pastry, but a loose pancake with oats or other stuff. The French invented apple baked with butter, and the British invented apple pie known to most people. This classic staple food includes crunchy pie base, crisp pie crust, soft apple flesh with proper roasting.

In the book *Harry Potter*, the apple pie that Harry tasted at the start-of-the-term banquet of the Hogwarts school is made in the most traditional and typical British style. It can be seen that the apple pie is still a common food in the UK.

During the Second World War, the war correspondents visited the American soldiers on the front line and asked, "Why did you go to war?" The soldiers who answered the questions were full of tears and said with deep affection: "For my mom and apple pie." Since then, apple pie has become the symbol of the United States.

State dinner for Chinese leaders was planned by former First Lady Michelle Obama and prepared by White House "first Chef". The main courses included boiled crayfish of Maine, dry-cooked ribbed beefsteak, onions cooked with skim milk, and black mushroom soup. Appetizers were cold cheese salad, potato and spinach cream sauce, while desserts were the most commonly seen apple pie in the United States, which proves that apple pie plays an important role in the banquet and is a typical American food.

The pie is a kind of dessert containing stuffing and pie pastry. It is generally round and the upper pastry is made like a net. The pie wrapper is golden, soft and crisp, and the stuffing is sweet and delicious.

派是由派馅和派皮两部分烤制而成的一种甜点，制作方法简单但营养丰富，深得欧美各国人民喜爱。因派馅与派皮种类不同，派的种类也各式各样。最经典的“苹果派”已成为美国人生活中常见的一种甜点。

一 派的起源及发展

最早出现的派为苹果派，起源于欧洲。五百年前，人们缺乏食物，便大面积种植苹果。因宗教原因认为生吃苹果意味着犯了禁忌，所以将苹果烤熟后当成主食用刀叉切着吃。烤制的苹果肉丰厚绵软，果糖因脱水浓缩而更加甜美，成为淀粉类食品的最佳替代。后来法国人在苹果底部开口、去芯，再填入黄油和砂糖。烤制之后，黄油融化，为烤苹果增加了类似肉类的丰腴，砂糖则增加了果肉的甜味。除了用黄油、砂糖之外，西班牙人还在烤苹果内加入红酒、坚果，甚至在烤苹果表面淋上饴糖和巧克力。北方寒冷地区的人们倾向于在烤苹果里加入淀粉，做成主食，这便就是最早的苹果派。

随着烤果盛行于南欧，苹果派在北欧也流传开来。瑞典是最早将苹果放进派的北欧国家，哥特人的派没有派底，上面的壳也不是一层均匀的编织状面饼，而是松散的、加入了燕麦或者其他粗面的“烙饼”。法国人发明了黄油烤苹果，英国人发明了大众意义上的苹果派。焦脆的派底、起酥的派皮、绵软的苹果肉和恰到好处的烘烤，成就了这道经典的主食。

在美国殖民地时期，由于粮食种植周期短，苹果树培育周期长，苹果不足而粮食有余，让思乡心切的美国人发明了美式苹果派。其所含苹果更少，面皮更厚，且在面皮上划几刀，使其在烘烤过程中透气，其特点是甜味减低，香味更浓。美国立国的国父们无不以美式苹果派作为国家自豪感的寄托，托马斯·杰斐逊有他的私家配方，本·富兰克林喜欢吃苹果派便在后院种起了苹果树，华盛顿夫人玛莎·华盛顿在外交宴请上必须有苹果派。至今，每年7月4日美国独立日，苹果派都是家家户户不可缺席的美味佳肴。

目前，随着时代的发展，诞生了各类形形色色的派，如柠檬派、柠檬塔、柠檬蛋白派、起司派等酸甜口味的派，蓝莓派、草莓派、樱桃派等水果派，是既好看又美味的甜点。欧洲和澳大利亚的派与美国派完全不同，作为早餐或简餐主食，一般是咸的，很多是肉馅的。

二 派的逸闻趣事

《哈利波特》一书中，哈利在霍格沃茨开学宴会上尝到的苹果派，就是最为传统和典型的英式做法。可见，苹果派在英国仍然是一种常见的食品。

第二次世界大战时，战地记者访问前线的美国士兵：“你们为什么要去打仗？”回答的士兵满含眼泪，深情地说：“For my mom and apple pie.（为了母亲和苹果派）”。

由美国前第一夫人米歇尔筹备、白宫“第一厨”康莫福德主持的款待我国重要领导人的国宴，开胃菜是梨子山羊乳干酪沙拉、马铃薯和菠菜奶油沙司；主菜为水煮缅因州大龙虾、干式熟成肋眼牛排配脱脂乳酥洋葱、黑蘑菇汤；甜点就是美国最常见的苹果派，足以证明苹果派在宴席上的地位。

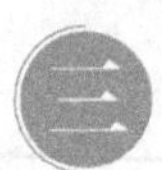

三 派的分类

从最初传统的苹果派开始，派经过了几百年的发展，现如今已种类繁多。

按照派皮的不同，可分为单层派（牛奶鸡蛋布丁派、南瓜派、奶油布丁派、布丁戚风派、冷冻戚风派）、双皮派（肉派、水果派）、油炸派（油炸苹果派、油炸樱桃派）；按派馅可分为肉派（牛肉派、鸡肉派等）、水果派（苹果派、樱桃派、菠萝派）等。

按馅料味道分为咸馅饼和甜馅饼两大类；按馅料填充方式可分为三类：一类是把派皮铺在烤盘上，接着倒入馅料；第二类是先将馅料倒入盘中，再铺上派皮；第三类是馅料完整地包覆在派皮里。

西方的派与中国的馅饼有些类似，都是烤焗食品，一般由饼皮包着馅料。馅料是各种类型的食品。例如，肉类、蔬菜、水果及蛋黄酱等，不同的是西方的派大都以甜为主，偏向于糕点类方向发展，而中国的馅饼大都以咸为主，偏向于主食方向发展。

四 派的制作工艺

（一）原辅料介绍（以苹果派为例）

低筋面粉

低筋面粉蛋白质含量低，麸质较少，因此筋性较弱，可保证制成的派皮松脆。

苹果

苹果性平，味甘酸甜，含有丰富的营养物质。苹果炒熟做派心，香甜可口，营养丰富。

黄油

用黄油翻炒苹果粒，可增加风味。

砂糖

砂糖可增加甜味，且细小颗粒的白砂糖更容易融入面团中。

肉桂粉

肉桂粉是由肉桂或大叶清化桂的干皮和枝皮制成的粉末，香味浓郁，主要作为天然香辛料用于食品的调味和抗菌防腐，因其富含钙、铁、锰等矿物元素，具有高的食用和药用价值，可增加派馅的风味。

蛋液

表面刷蛋液后可以使派在烘焙后产生明亮金黄的色泽，既使派品相美观又增加食欲，也使派更具风味。

（二）派的制作

派制作工艺流程

和面→冷藏→制馅→擀皮→铺盘→加馅→切条→编制→刷蛋液→烘烤

参考配方

面团 / 派皮：低筋面粉 450g，黄油 180g，水 150g，细砂糖 30g。

苹果馅：苹果 800g，砂糖 90g，水 60g，玉米淀粉 30g，柠檬汁 30ml，蛋液 15ml，肉桂粉 6g，食盐 5g。

操作要点

1．和面：黄油室温软化，至可以用手指按压下去；软化好的黄油中加入面粉和细砂糖，将黄油、面粉、糖抓匀并搓成颗粒状，再加入清水揉成面团，这样可以确保派皮更均匀。派皮面团不能揉太长时间，成团即可。

2．冷藏：派皮面团用保鲜膜包好放在冰箱里冷藏 1h 左右，备用。

3．制馅：将苹果洗净、去皮、去核，并切成小块，用少许黄油翻炒，翻炒几分钟后加入细砂糖继续翻炒，苹果翻炒至变软时将调好的玉米淀粉汁倒入锅中继续翻炒，直至出现黏糊状，再加入食盐、肉桂粉、柠檬汁等，搅拌均匀，放凉备用。

4．擀皮：从冰箱里取出派皮面团，分成两部分（重量 1∶2），其中 1/3 作派底，用擀面杖将两部分分别擀成薄片（约 5mm）。

5．铺盘：把擀好的派皮铺在派盘上，轻压派皮，使派皮紧贴在派盘上，用擀面杖在派盘上滚过，将多余的派皮切断，移去多余的派皮。用叉子在底部的派皮上均匀地叉上一些小孔。

6．加馅：将馅倒入派皮中，一般为九分满，不能太满，以防止烘烤后馅皮破裂。

7．切条：将擀好的另一张派皮用刀切成均匀的小长条。

8．编制、刷蛋液：切好的细条交叉地编好在派盘上，压去多余的面条，刷上蛋液。

9．烘烤：烤箱200℃预热后放入苹果派烘烤15min后，烤箱温度改为170℃烘烤25min左右即可。

五 派的风味特色

派是烤制而成的一种甜点，一般呈圆盘状，上层派皮编织成网。派皮色泽金黄均匀，糕坯断面为淡黄色；外形整齐，无不规则大空洞，无糖粒、粉块，馅料分布均匀；无异味，口感酥软松脆，派馅香甜可口，细腻润滑，无明显沙粒感，具有独特风味（肉桂粉特殊的味道）。

苹果派富含碳水化合物，能够为机体提供必要的能量，可作为主食。派馅富含维生素、多糖及矿物质元素，具有一定的营养价值。

参考文献

董青，刘斐．2010．鲜奶酪巧克力派的加工工艺研究［J］．中国乳业，（08）：74-76．
董艳华．2016．肉桂苹果派罐头的研制［J］．科技展望，26（27）：68．
方献群，黄旭文．2004．蛋黄派的生产技术［J］．中小企业科技，（01）：24-25．
郭桦，陈朝毅，郭祀远．2006．食品添加剂在蛋黄派生产中的应用［J］．中国食品添加剂，（04）：144-146．
郭娇娇，黄秀娟，陈日春，等．2016．低卡休闲蛋黄派配方的优化［J］．农产品加工，（15）：21-25．
贾文婷，张萍，王利敏，等．2015．一种油炸蛋黄派的加工方法［J］．肉类工业，（06）：7，10．
梁永平，刘圣本，何力，等．2016．正交实验优化冻干香蕉苹果派的配方和工艺［J］．轻工科技，（02）：12-13，48．